ScïencePlus®
Interactive Explorations

CD-ROM for Macintosh® and Windows®

Teacher's Guide
Level Green

HOLT, RINEHART AND WINSTON
Harcourt Brace & Company

Austin • New York • Orlando • Atlanta • San Francisco • Boston • Dallas • Toronto • London

To the Teacher

Imagine having access to a fully equipped laboratory where your students could study questions and problems related to life processes, geology, energy and power, and chemical reactions. This is exactly what is possible with *SciencePlus Interactive Explorations*. By using this innovative CD-ROM program, students get valuable experience in a unique laboratory setting.

Welcome to Dr. Crystal Labcoat's laboratory, where with the click of a mouse, students will have access to a variety of scientific tools and equipment and will be challenged to solve some perplexing problems and mysteries. Dr. Labcoat operates a virtual laboratory, and your students are her lab assistants. Under her guidance, your students will perform some amazing and highly interesting scientific experiments and studies. But be prepared—although Dr. Labcoat provides the lab and the equipment, your students provide the brainpower.

This *Teacher's Guide* consists of the following components:

- **User's Guide**
  The User's Guide provides important technical information about the program, including its installation, features, and use.

- **Teaching Notes, Worksheets, and Handouts**
  Organized by exploration, this information includes background material and worksheets that guide students through the CD-ROM experience and allow them to record their answers on paper, rather than electronically. In addition, the Computer Database articles (also called CD-ROM articles) are provided so that they can be used as handouts. The worksheets and handouts make the program more flexible in cooperative groups or when computer time is limited.

- **Answer Keys**
  Worksheet pages with overprinted answers are provided for each exploration to make grading worksheets and fax forms fast and efficient.

Copyright © by Holt, Rinehart and Winston.

All rights reserved. No part of this publication may be reproduced or transmitted in any form or by any means, electronic or mechanical, including photocopy, recording, or any information storage and retrieval system, without permission in writing from the publisher.

Permission is hereby granted to reproduce Blackline Masters in this publication in complete pages for instructional use and not for resale by any teacher using SCIENCEPLUS INTERACTIVE EXPLORATIONS.

Photo/Art Credits
All work, unless otherwise noted, contributed by Holt, Rinehart and Winston.
Front cover (fish image): Tom McHugh/Steinhart Aquarium/Photo Researchers, Inc.

SCIENCEPLUS is a registered trademark licensed to Holt, Rinehart and Winston.

Printed in the United States of America

ISBN 0-03-018534-3   3 4 5 6 7 8 9 021 00 99

Contents

User's Guide vi
- System Requirements vi
- Installing the Program vii
- Logging On viii
- Welcome to the Main Menu ix
- Using the Virtual Laboratory x
- Assessment Tools xiii
- Networking Student Reports xvi
- Optimizing Performance xvii
- Technical Support Information xviii

Exploration 1: Something's Fishy 1

African Cichlids are dying at the local fish store. The store's manager, Ray McMullet, needs to know what changes to make in order to save the fish and prevent this trauma at the Fishorama.

Exploration 2: Shut Your Trap! 9

Plant poachers are threatening the existence of Venus' flytraps in the wild. Lily N. Lotus wants to find out what the optimal growth conditions are for these carnivorous plants so that they may be easily grown in nurseries.

Exploration 3: Scope It Out! 18

Some microorganisms have been extracted from the digestive tract of an ancient, amber-entombed bee. Dr. Viola Russ wants to classify these gutsy microorganisms and explain the role that they might have played in the bee's life.

Exploration 4: What's the Matter? 27

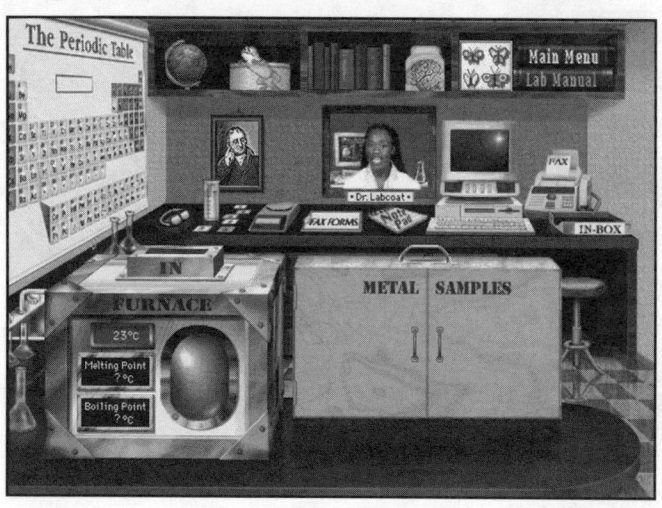

Scientists in Hawaii are in a hot spot—the tip of one of their instruments, the lava analyzer, has melted. Dr. John Stokes and his team need to know the most practical metal to use to replace the tip of the lava analyzer.

Exploration 5: Element of Surprise 38

Fred Stamp is packing some samples of chemical elements and shipping them to the South Pole. He needs to ensure that the shipment reaches a remote research station safely without encountering any explosive surprises.

iv   SCIENCEPLUS INTERACTIVE EXPLORATIONS TEACHER'S GUIDE • LEVEL GREEN

Exploration 6: The Generation Gap 48

Wendy Powers operates a small log-home manufacturing plant called Eco Cabin, Inc. She wants to know if the Electroprop, a wind turbine, would help her customers generate some savings on their electricity bills.

Exploration 7: Teach It While It's Hot! 58

Mr. McCool is teaching a lesson on the relationship between temperature and heat to his class of middle-school students. He wants to make sure that this lesson is cool enough to get his students fired up about temperature and heat.

Exploration 8: Flood Bank 68

A local environmental-impact committee is debating a reservoir of pros and cons about whether to build a dam on a nearby river. The committee chairperson, Sandy Banks, wants to find out what impact the dam would have on the local river environment.

Answer Keys 77

User's Guide

SYSTEM REQUIREMENTS

Before you begin using *ScientificPlus Interactive Explorations,* you will need to acquire the necessary equipment and set it up properly. The complete setup includes a computer (IBM®-compatible or Macintosh®-compatible) connected to a CD-ROM drive. Audio headphones are optional. To use the program with a network, you will also need additional cables to connect the machines, a dedicated network file server, and a network operating system.

A note concerning minimum requirements: *Although the program will run on machines with the minimum requirements listed below, we strongly recommend that the program be used on newer model computers (040 Macintoshes and 486 PCs or higher) that are more efficient at handling the demands of multimedia. If you run the program on lower-end machines, you may experience slow response times to mouse clicks, longer loading times for video and sound, and slower animations, as well as occasional dropped video frames and sound.*

COMPUTERS

Macintosh®-Compatible Computers

Minimum Requirements:
- 68030 CPU running at 25 MHz or higher **(Highly recommended: 68040 CPU running at 20 MHz or higher)**
- 13-inch or larger color monitor capable of displaying 256 colors at 640 × 480 resolution
- System 7.1 or higher
- Double-speed or higher CD-ROM drive
- 8 MB of RAM
- 30–40 MB of free memory on hard drive if you plan to install individual explorations (explorations can also be run directly from the CD-ROM)
- QuickTime® for Macintosh® (provided with the program)
- Internal/external speaker(s); headphones (recommended for classroom settings)

IBM®-Compatible Computers

Minimum Requirements:
- 80386 DX running at 25 MHz or higher **(Highly recommended: 80486 running at 33 MHz or higher)**
- 13-inch or larger color monitor capable of displaying 256 colors at 640 × 480 resolution
- Windows® 3.1 or higher
- Double-speed or higher CD-ROM drive
- 8 MB of RAM
- 30–40 MB of free memory on hard drive if you plan to install individual explorations (explorations can also be run directly from the CD-ROM)
- QuickTime® for Windows® (provided with the program)
- Sound Blaster™ or other compatible sound card
- Internal/external speaker(s); headphones (recommended for classroom settings)

PRINTERS

Minimum Requirements:
- Laser printer, ink-jet printer, or 24-pin dot-matrix printer

INSTALLING THE PROGRAM

... ON MACINTOSH®-COMPATIBLE COMPUTERS

1. Place the CD-ROM in the CD-ROM drive.
2. A window will appear with the program's "Read Me" file and the **SciPlus Level Green Installer** icon. Double-click this icon and follow the procedures on the screen.
3. For information concerning installation of individual explorations, please see the "Read Me" document on this screen.
4. After installation is complete, a **SciencePlus Green** folder will appear on your hard drive. Open the folder and double-click the **SciencePlus Green** icon to launch the program.

... ON IBM®-COMPATIBLE COMPUTERS (WITH WINDOWS®)

1. Place the CD-ROM in the CD-ROM drive.
2. Locate the **Install** and double-click it.
3. Once the program is installed, a window will appear with the program's "Read Me" file, the **SciPlus Green** icon, and the **Uninstall SP Green** icon.
4. Launch the program by clicking the **SciPlus Green** icon found in the directory or folder of Program Manager.

Note: *For details on how to do a custom installation of individual units to improve performance, please see the "Read Me" file located on the CD-ROM.*

LOGGING ON

Logging on to *SciencePlus Interactive Explorations* is quick and simple. After launching the program, a log-on display will ask your students if they are working as guests, as individuals, or as a group.

... AS AN INDIVIDUAL

To log on as an individual, a student follows this procedure:

1. The student clicks the **Individual** button. A dialog box will appear asking the student to type in his or her first and last name, and the teacher's name and class or period.
2. The student clicks **Enter** or presses **Return.**
3. When the main menu appears, the student chooses an exploration.

... AS A GROUP

To log on as a group, students follow this procedure:

1. Students click the **Group** button. Students are then asked to type in their one-word group name, the teacher's name and the class or period, and the names of the members of their group.
2. Students click **Enter** or press **Return.**
3. When the main menu appears, the group chooses an exploration.

... AS A GUEST (Nonassessed Use of the Program)

The log-on contains a guest feature that allows you or your students to do an exploration without engaging the assessment function of the program. In other words, what is completed using the guest feature will not be graded. To log on as a guest, a student clicks the **Guest** button. When the main menu appears, the student chooses an exploration.

An Important Note: *Please be sure to consult the "Read Me" file found on the CD-ROM. There you will find important, up-to-date information concerning general technical issues and changes that may not be present in this guide. Likewise, an additional "Read Me" file is contained in the Assessment Tools folder of the CD-ROM. This file identifies any updates for using the assessment tools.*

WELCOME TO THE MAIN MENU

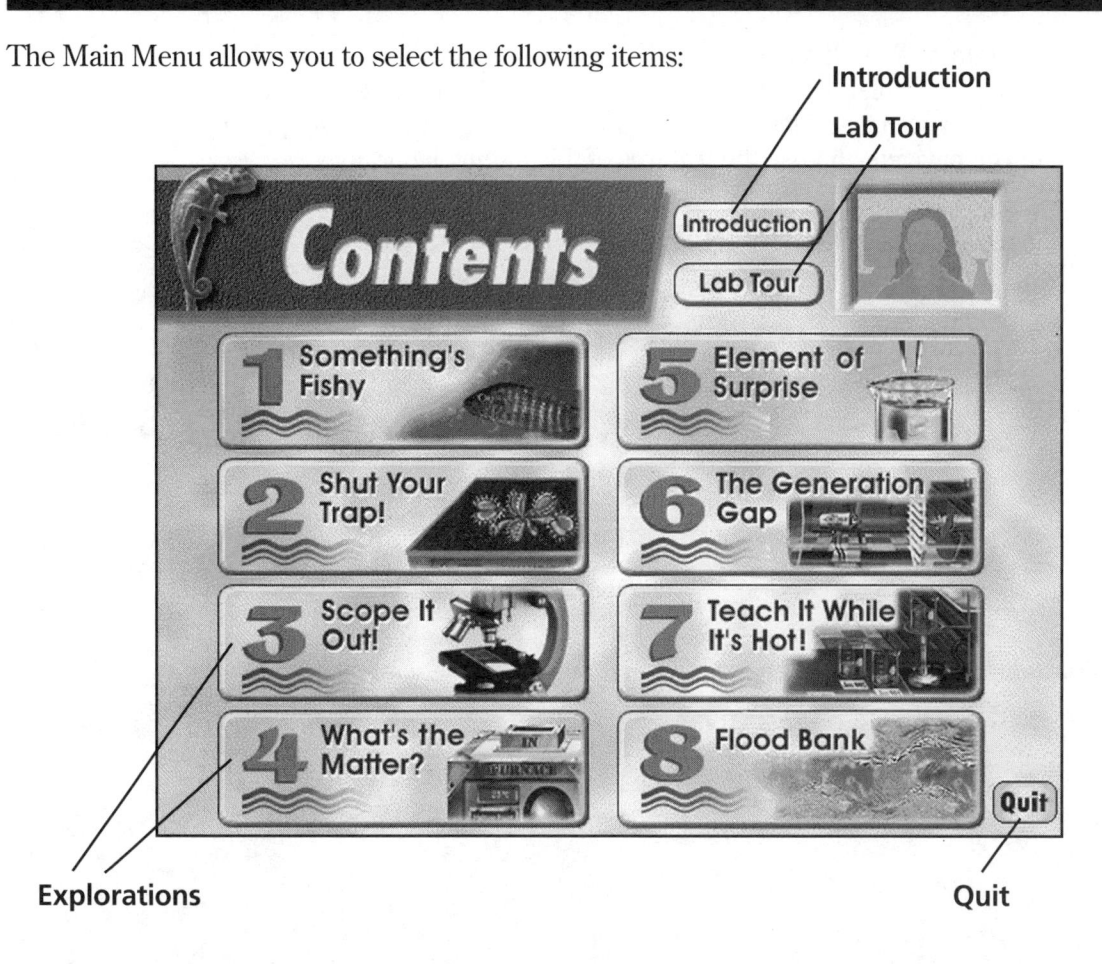

The Main Menu allows you to select the following items:

Explorations — Introduction, Lab Tour, Quit

- **Introduction**

  Click the **Introduction** button to get a general overview of the explorations.

- **Lab Tour**

  Click the **Lab Tour** button to get a quick yet comprehensive tour of Dr. Crystal Labcoat's laboratory. This is an excellent way for both you and your students to get acquainted with the standard equipment and features of the lab. Of course, depending on the exploration and the problem to be solved, Dr. Labcoat has a wide variety of specialized equipment, which is described in the exploration in which the equipment is used.

- **Interactive Explorations**

  Simply click any one of the eight **Exploration** buttons to start that particular exploration. Should you want a quick overview of an exploration, move the cursor over any exploration button, and a brief description of the exploration will automatically appear on the screen.

- **Quit**

  You can exit the program by clicking the **Quit** button on the main menu. Within an exploration, you can also end your session by selecting **Quit** in the pull-down menu under **File**.

USER'S GUIDE   ix

USING THE VIRTUAL LABORATORY

Dr. Labcoat's laboratory is a rich and functional scientific setting where students can practice their scientific problem-solving skills as well as their process skills as they try to solve a variety of science-related problems and mysteries. The following information will help you navigate and use the features of this unique laboratory.

This virtual laboratory is from Unit 1, Something's Fishy.

NAVIGATION

Navigation is accomplished by moving the mouse, which operates the cursor. You will notice that four types of cursors are used in the explorations.

Arrow Cursor	This point-and-click cursor is used to close pop-up windows as well as to set variables in experiments that require adjustments before a simulation can begin.
Pointing-Finger Cursor	This point-and-click cursor indicates the areas of the lab that are active. Activate a selection by clicking it.
Hand Cursor	This cursor indicates movable objects. To move objects, click and hold the mouse. You will be able to drag objects to lab equipment for testing as well as for classification and storage.
Bar Cursor	This point-and-click cursor appears in fields that require typing or word processing.

x   SCIENCEPLUS INTERACTIVE EXPLORATIONS TEACHER'S GUIDE • LEVEL GREEN

TOOL BAR

A tool bar featuring pull-down menus will display the following items:

File	Edit	Sound	Windows	Options
Quit	Copy Select All **Active when you are copying text from the Computer Database**	Level 0 Level 1 Level 2 Level 3 Level 4 Level 5 Level 6 Level 7	Notepad **Access to other pop-up variable panels available in certain explorations**	English Audio Spanish Audio **Available for the Lab Manual only**

LAB MANUAL

The Lab Manual provides a short explanation of the purpose and the operation of each piece of equipment in the lab. This is a handy reference for students who may need additional help.

To page through the Lab Manual, click the tabs at the lower right corner of the page. You can play audio instructions in English by clicking the megaphone-shaped icon. To hear instructions in Spanish, click **Options** on the tool bar and then select **Spanish Audio.** To hide the Lab Manual, click the **Close** box in the upper right corner.

FAX MACHINE

The Fax Machine is a primary means of communication both to and from the lab. Incoming fax messages identify problems to be solved. Outgoing fax messages are generated by students as they solve the problems and are asked to communicate their findings. Faxes often contain many pages. To page through faxes, click the tabs at the lower right corner of the page. To hide faxes, click the **Close** box in the upper right corner.

IN-BOX

The In-Box is where fax messages and other correspondence are kept for reference at any time during the exploration. To page through faxes and other correspondence, click the tabs at the lower right corner of the page. To hide the correspondence, click the **Close** box in the upper right corner.

NOTEPAD

The Notepad is always available for jotting down notes and observations. Simply click the Notepad and start taking notes. Students can even paste articles from the Computer Database into the Notepad. Since the contents of the Notepad are not saved by the program, students should print their Notepads before leaving an exploration if they are interested in keeping a record of their notes.

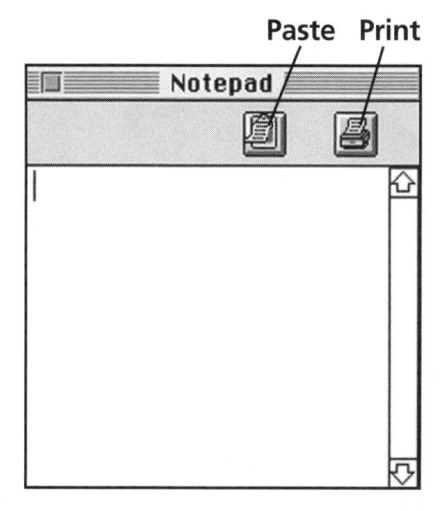

To paste text from the Computer Database into the Notepad, simply highlight the desired passages (or click **Select All**) and then select **Copy** from the pulldown **Edit** menu. Go to the Notepad and click the **Paste** icon. The text that you copied from the Computer Database will appear in the Notepad. **To print the contents** of the Notepad, simply click the **Print** icon.

More advanced students may choose to take notes or paste information from the Computer Database into other documents (such as those created by SimpleText) by running a separate word-processing application. Although this method is more complex, it allows students to save their notes electronically. For information concerning the simultaneous use of two applications, refer to the user's guide that accompanies your system's software.

COMPUTER DATABASE

Students can use this Computer Database to easily access information on a variety of subjects. The information is organized into articles that contain text as well as illustrations, photographs, and video. The Computer Database contains information that is vital to solving the problem or mystery.

To access articles that are relevant to an exploration at hand, students simply scroll to the applicable topic and click on subtopics to view that information. Students can also access the complete database of articles by clicking the **Database Index** button. To return to the exploration from which they accessed the database, students click **Table of Contents.**

It is important to note that students can copy text from the Computer Database into their notepads or other word-processing documents for printing. Images and video appearing in the Computer Database, however, cannot be pasted into the Notepad.

FAX FORMS

Fax forms located on the clipboard contain the forms necessary to create a new fax. Sending a fax is how students will communicate their solutions to the problems. Once a fax form has been completed, it is sent by pressing the **Send It** button. An appropriate response will be forthcoming from both the requester of the information and from Dr. Labcoat.

CALCULATOR

In explorations that require mathematical calculations, a calculator is provided in the lab. Simply click the calculator to bring up the calculator's keypad.

ASSESSMENT TOOLS

A variety of assessment tools are available to make grading and record keeping as simple as possible. The assessment tools are located in the following folders, depending on which CD-ROM level (Green, Red, or Blue) you are using.

SciencePlus Level	Folder Name
Level Green	**adminSPG**
Level Red	**adminSPR**
Level Blue	**adminSPB**

If you are using a Macintosh®, you can access the Assessment Tools folder by entering the Preferences folder of your System folder. **If you are using a PC,** you will find the Assessment Tools folder in the Windows® folder. You can also do a search for the Assessment Tools folder by entering the folder name as identified in the chart above.

When you open an Assessment Tools folder, you will find the following folders:

- Read Me
- Student Reports
- Answer Keys
- File Management

To view the contents of any folder, simply double-click the folder.

READ ME FOLDER

All updates to the assessment tools component of the program can be found in this printable file.

STUDENT REPORTS FOLDER

You can access student reports after each class session or at the end of the day by double-clicking the folder labeled "Student Reports." In the folder, you will see your students' work in the form of SimpleText files. These documents can be opened in SimpleText or in another word-processing application, if you choose.

Each student file or record has a prefix consisting of up to five letters. In the case of individual students, the prefix consists of the first four letters of the student's last name and the first initial of the student's first name; in the case of groups, the group's name (up to five letters) will appear. The prefix is followed by a period (.) and a three-letter suffix (indicating the level of *SciencePlus Interactive Explorations* and the exploration number).

Consider the following examples:

File Name	Individual or Group Name	Level	Exploration
Smitj.GR1	Smith, Jennifer	Green	1
Smitp.GR2	Smith, Peter	Green	2
Aces.RD7	Aces	Red	7
Diamo.BL8	Diamonds	Blue	8

All student records relate directly to the fax forms of the *ScancePlus Interactive Explorations,* where students produce their work. Each student report contains the following sections:

Student/Class Information Section
This section contains:
- Student name or group name with names of group members
- Teacher's name
- Class or period
- The date the exploration was conducted
- The duration of the student's work within the exploration

Computer-Graded Section
This section contains the student's responses to the close-ended questions on the fax form. These questions, which are indicated with an asterisk (*), will have already been graded by the computer. Fifty (50) points are possible.

Teacher-Graded Section
This section consists of the answers to the open-ended questions on the fax form. This section must be read and evaluated by the teacher. Fifty (50) points are possible.

Scoring Section
This section consists of three scores: the computer-graded score, the teacher-graded score, and the composite score (final grade). Suggestions for grading student work can be found in the Answer Keys folder.

Teacher Comments Section
This section provides a space for making comments on your students' work. Simply type in your comments as you view the Student Report, or you can print out the Student Report and write in your comments manually.

ANSWER KEYS FOLDER
The Answer Keys folder contains eight answer keys, which correspond to the eight explorations for a particular level. The files are named as indicated in the chart below. Please note that this chart only shows the file names for Exploration 1 in levels Green, Red, and Blue. For Exploration 2, the file names would end in "2."

File Name	Level	Exploration
AnkeyGR1	Green	1
AnkeyRD1	Red	1
AnkeyBL1	Blue	1

If desired, you can customize the names of these files simply by renaming them. Also, if you have installed Level Red or Blue on the same machine, their respective contents will be found in another folder on your hard drive.

To use an answer key:

1. Double-click the answer key you wish to view.
2. Resize the answer key if you would like to view it on-screen along with a student report.
3. Grade your students' responses to the open-ended questions on the student report. Remember, the computer-graded questions are scored automatically.
4. If desired, print out hard copies of the answer key, the student's report, or both.

FILE MANAGEMENT FOLDER

The File Management folder contains an example of how you might set up class folders for storing your files. You can use this simple folder structure by changing the names of folders or adding more folders to suit your needs.

We suggest that you set up your records by class. At the end of each class, move student files into their corresponding class folder. You may decide to set up folders within each class folder by student or by exploration number. Viewing files by date and time will be helpful in determining the class folder into which a student file can be inserted.

If you need help structuring your folders, please consult your computer manual for more in-depth instructions.

NETWORKING STUDENT REPORTS

The student reports in the Assessment Tools folder can be networked. This allows you to gather student reports to a central computer. To do this, you will need a dedicated network file server meeting the following requirements:

Macintosh®-Compatible Computers

Minimum Requirements:
- 68040 CPU or higher, such as a Quadra or PowerPC
- 16 MB of RAM
- Recommended: 20 K per user per semester (or 40 K per user per year)

IBM®-Compatible Computers

Minimum Requirements:
- Minimum 486DX running at 33 MHz or higher
- 16 MB of RAM
- Recommended: 20 K per user per semester (or 40 K per user per year)

You will also need Ethernet cards and cables, Novell Netware® 3.0 or higher, Windows® NT ™ 3.51 or higher, or AppleShare® 3.0 or higher network operating system.

OPTIMIZING PERFORMANCE

There are many things you can do to optimize the performance of *ScoiencePlus Interactive Explorations*. Consider the following list of options:

1. Make sure that your monitor's resolution is set to 256 colors. Running the program on "Thousands of Colors" may slow the program down.

2. Do a custom installation of an exploration rather than running the program completely off the CD-ROM. The program will perform better if an exploration is installed onto your hard drive. Remove explorations from the hard drive when you move on to a new exploration.

3. If your hard drive is more than 80 percent full, performance may suffer. Remove old files and applications that are no longer pertinent or useful. Consider storing them on another hard drive or with alternative methods of data storage (such as Zip™ cartridges).

4. Use a utility to optimize your hard drive. Refer to the manual that came with your computer for more information about optimizing your hard drive.

5. If you are running other applications in the background—no matter how simple or complex the application—performance will suffer. Quit all applications other than *ScoiencePlus Interactive Explorations*.

6. If you are on a network or on the Internet, your chances of experiencing freezes, crashes, and poor video/audio are much greater than if you are not.

7. Make sure you have the most current version of QuickTime 2.1 and Quicktime Powerplug extensions and the latest version of Sound Manager extension (v. 3.2). Using older versions of these extensions may result in reduced video quality.

8. If you have been running any memory-intensive applications prior to running *ScoiencePlus Interactive Explorations*, restart your machine before running the program.

9. If the total memory requirement of your system software and this program approaches the limits of your machine's total RAM, performance will suffer. Consider upgrading your machine's capability by adding more RAM. Also consider using an extensions-management utility to turn off extensions not being used by the application or temporarily placing unused extensions in a folder labeled "Disabled Extensions."

10. Turn off your computer's Virtual Memory.

Suggestion: *If students are working in a setting that includes lower-end machines as well as newer models, you may want to rotate students or groups from one machine to another. Have your students do their research with the Computer Database section of the program on lower-end machines and do the main experiments using the better performing machines.*

TECHNICAL SUPPORT INFORMATION

At Holt, Rinehart and Winston we recognize the importance of providing you with the answers and help you need to use our quality instructional-technology products to their fullest potential.

Because systems, technology, and content are often inseparable, HRW has assembled a team of dedicated technical and teaching professionals and a suite of comprehensive support services to provide you with the support you deserve, 24 hours a day, seven days a week.

Technical Support Line (800) 323-9239

The HRW Technical Support Line, which operates from 7 A.M. to 6 P.M. Central Standard Time, Monday through Friday, puts you in touch with trained Support Analysts who can assist you with technical and instructional questions on all of HRW's instructional technology products.

Technical Support on the World Wide Web   http://www.hrwtechsupport.com

Contact the HRW Technical Support Center 7 days a week, 24 hours a day, at our site on the World Wide Web. Simply select the product you are interested in, and with a click of the mouse you can receive comprehensive solutions documents, answers to the most frequently asked questions, product specifications and technical requirements, and program updates from our FTP site. You can also contact our analysts at the Support Center using the following E-mail address: **tsc@hrwtechsupport.com**

Technical Support via Fax (800) 352-1680

Get the solutions you need with the HRW Technical Support Center's fax-on-demand service. Simply give us a call at our toll-free number to receive product-specific solutions within minutes. Our fax-on-demand service is available 7 days a week, 24 hours a day.

Exploration 1
Teacher's Notes

Something's Fishy

Key Concepts	A controlled experiment is an efficient way of determining how individual environmental variables affect living things. Acidic water conditions affect the health of African Cichlids.
Summary	African Cichlids are dying at the Fishorama store. The manager, Mr. McMullet, needs to know what changes he should make in his tanks to save the fish.
Mission	Find out why the African Cichlids are dying.
Solution	The pieces of ornamental driftwood that Mr. McMullet has been using in his aquariums have been leaching tannins into the water. Tannins are compounds that make the water more acidic. Such a change can be devastating to the health of the organisms that live in the water. The African Cichlids have been dying as a result of the water's increased acidity.
Background	Although difficult to care for, cichlids are interesting aquarium fish because of their bright colors. Unlike many other fishes, which abandon their offspring at birth, cichlids protect their young. Many aquarists enjoy watching the adult fish accompany the young cichlids around the tank.
	Scientists find the cichlids of African lakes particularly interesting because they speciate (develop into new species) so rapidly. In Lake Victoria, as many as 300 different species of cichlids have evolved since the lake formed 750,000 years ago. Such rapid evolution generally takes millions of years, and scientists are investigating environmental factors that could be causing this evolutionary explosion. Scientists are also concerned about how cichlid species are affected by human activities. Because many cichlid species are confined to lakes, they are especially vulnerable to environmental changes. The more we understand about the environmental changes that affect cichlids, the better our chances of ensuring their survival.

Exploration 1 Teacher's Notes, continued

Teaching Strategies

One purpose of this Exploration is for students to discover how to perform a controlled experiment effectively. Ideally, students will learn to do so by examining the CD-ROM articles provided and by analyzing their own successes and mistakes with the experiment. If necessary, review the CD-ROM articles about the scientific method with students who require additional assistance. Make sure students understand that observing a change in the experimental tank is only part of their goal. Isolating the specific variable responsible for that change is the most important part of their task. By comparing results from the experimental tank with results from the control tank, the effect of the variable can be determined. Changing the control tank in addition to the experimental tank would make it impossible to determine which variable is responsible for the results.

As an extension of this Exploration, you might wish to have students research the effects that changes in pH levels can have on species in lakes, streams, and rivers. Encourage students to find out more about the effects of acid precipitation on aquatic environments and to explore what is being done to protect wildlife in those areas.

Bibliography for Teachers

Ciresi, Rita. "One Fish, Two Fish, Red Fish, Blue Fish." *Penn State Agriculture*, Winter 1990, pp. 33–35.

Morris, Ronald. *Acid Toxicity and Aquatic Animals*. Cambridge, MA: Cambridge University Press, 1989.

Wilson, Edward O. *The Diversity of Life*. Cambridge, MA: Belknap Press of Harvard University Press, 1992.

Bibliography for Students

Encyclopedia of Aquatic Life. Keith Banister and Andrew Campbell, ed. New York, NY: Facts on File, 1985.

Kluger, Jeffrey. "Go Fish." *Discover*, 7 (18): March 1992, p. 18.

Other Media

Cichlid CD-ROM
Infobase Press
1844 S. Columbia Lane
Orem, UT 84058
801-221-1117

In addition to the above CD-ROM, students may find relevant information about cichlids on the Internet. Interested students can search for articles with keywords such as the following: *cichlids*, *aquariums*, *pets*, *hobbies*, and *tropical fish*.

Note: Remind any students who are interested in adding cichlids to their home aquariums that cichlids are difficult to care for. Encourage students to carefully research the special needs of cichlids before attempting to add them to their aquariums.

Name _____ Date _____ Class _____

Exploration 1 Worksheet

Something's Fishy

1. What kinds of problems is Mr. McMullet having with his African Cichlids?

   _____
   _____
   _____
   _____

2. What are five variables that might be affecting the African Cichlids?

   a. _____
   b. _____
   c. _____
   d. _____
   e. _____

3. What will you use for a control as you conduct your investigations?

   _____
   _____

4. Why is this control necessary?

   _____
   _____
   _____

5. Would it be better to test one variable at a time or several variables at once? Why?

   _____
   _____
   _____

6. Form a hypothesis for each of the experiments you conduct.

   Hypothesis 1: _____
   _____

Name _____ Date _____ Class _____

Exploration 1 Worksheet, continued

Hypothesis 2: _____
_____

Hypothesis 3: _____
_____

Hypothesis 4: _____
_____

Hypothesis 5: _____
_____

7. Record your observations as you investigate each hypothesis.

Hypothesis	Observations
1	
2	
3	
4	
5	

8. Were your experiments faulty in any way? If so, what steps did you take to correct your experiments?

_____
_____
_____
_____

Record your conclusions in the fax to Mr. McMullet.

4 SCIENCEPLUS INTERACTIVE EXPLORATIONS TEACHER'S GUIDE • LEVEL GREEN

Name _____ Date _____ Class _____

Exploration 1
Fax Form

FAX

To: Mr. Ray McMullet (FAX 512-555-8633)

From:

Date:

Subject: African Cichlids

What is your recommendation? _____
_____

✂·

For Internal Use Only

Please answer the following questions for my laboratory records. Scientists must always keep good records. *Dr. Crystal Labcoat*

During your experiments, which ONE of the following changes had a positive effect on the fish?

EXPERIMENTAL VARIABLES

☐ FEED FISH   ☐ INCREASE TEMPERATURE

☐ TURN LIGHT OFF   ☐ CHANGE FILTER

☐ REMOVE ORNAMENTAL DRIFTWOOD

Please explain why the African Cichlids responded to the above change.

_____
_____
_____
_____

What effect did the change that you made have on the fish?

_____
_____
_____

EXPLORATION 1 • SOMETHING'S FISHY

Exploration 1
CD-ROM Articles

Something's Fishy

The following articles can also be found by accessing the computer graphic of the CD-ROM for Exploration 1:

- *Freshwater Aquariums*
- *Fish for Freshwater Aquariums*
- *Controlled Experiments*

Freshwater Aquariums

Keeping tropical fish is a way to learn about a wide variety of fish species and their habitats. Aquariums provide a small ecosystem that is not only rewarding to create but also enjoyable to observe.

Hardware

The hardware needed to keep freshwater tropical fish is relatively simple and inexpensive. Basically, all you need is a tank, a filter system, and a water heater. In most cases, new hardware for a 10-gallon tank will cost around $60 to $75.

Tank—Aquarium tanks range in size from small 5-gallon tanks to large tanks that contain 150 gallons or more. Most beginners start with a 10- to 20-gallon tank. The tank should be made specifically for tropical fish and should be equipped with a lid with a light.

Filter System—Healthy fish require water that is free of visible particles, dissolved carbon dioxide, and excess ammonia. This is achieved in an aquarium with a filter system. Box-type filters are designed to circulate the water through a layer of spun glass and activated charcoal to clean the water. Biological filters rely on colonies of bacteria to keep the water clean. The most common biological filter is the under-gravel filter. This filter is little more than a grate placed underneath the gravel at the bottom of the tank. In operation, water is pulled through the gravel into the grate and then up through tubes at each corner of the tank. Colonies of bacteria living in the gravel break down the contaminants in the water.

Many box-type filters and all under-gravel filters require an air pump to circulate the water through the filter system. The air pump also helps add oxygen to the water.

Water Heater—Because most tropical fish come from areas near the equator, they require warm water with little temperature variation. A water heater is used to ensure that the aquarium water stays at a warm, uniform temperature.

Healthy Water for Fish

In most cases, freshwater aquariums can be filled with tap water. But the water must be treated first to remove chlorine and chloramine. These chemicals are added to municipal water supplies to kill harmful microorganisms, but these chemicals can also kill tropical fish. Inexpensive chemicals that will effectively dechlorinate water are available at pet stores.

The water should also be kept at a stable pH level. The pH level indicates the number of hydrogen ions in the solution. The pH scale ranges from 0 to 14. Acids, such as vinegar and lemon juice, have a pH below 7. Bases, such as ammonia, have a pH above 7. Most freshwater fish can survive a pH range of 6.5 to 7.5. Most tap water falls within this range.

The pH of your aquarium water can change, however. The waste products from fish are high in ammonia and can quickly alter the pH of the water if the filter system is not functioning properly. Having too many fish in the tank or feeding them too much food can also cause excess ammonia and a high pH. In addition, some decorations can make tank water more acidic. Coral, shells, or bits of limestone can raise the pH. Ornamental driftwood may leach plant tannins into the water and lower the water's pH.

Test papers are compared with color charts for pH readings.

Preparing the Water for Fish

If a biological filter is used, the tank should be set up and filled with dechlorinated water 2–4 weeks before adding any more than one or two hardy fish. This will allow time for bacteria to colonize the gravel at the bottom of the tank. Two species of bacteria are necessary: one species that converts ammonia to nitrites, and another species that converts the nitrites to relatively harmless nitrates. After a few weeks, the water should have a high level of nitrates and little or no ammonia. Additional fish can then be added.

Water is tested for ammonia, nitrites, and nitrates since high levels of these substances can harm fish.

Fish for Freshwater Aquariums

General Health

Before purchasing any fish, inspect them carefully. Make sure they do not have unusual spots, raised and flaking scales, or other signs of disease or parasites. A 10-gallon tank can hold about 20 small- to medium-sized fish. To reduce stress on the fish, add only 3–5 fish to the tank at any one time. Give these fish 1–2 weeks to adjust to the tank. It normally takes about 2 months to completely stock a new aquarium with healthy fish.

Tetras

Tetras are very hardy fish. These fish come from Africa, Central America, and South America. Tetras prefer to live in schools of at least 6 fish. Several species require acidic water (pH about 6.5), so you should consider the pH of your tap water before you invest in these fish. If you want to keep tetras but your water is not acidic, you might try the black tetra or the flame tetra.

Cichlids

Native to Africa, Asia, Central America, and South America, the cichlids (SIK-lids) are popular fish among aquarium owners. African Cichlids are extremely territorial and require places to hide, such as among plants, rocks, and wood. They also require water that is alkaline. Some cichlids live comfortably with other fish species, but other cichlids may eat other fish in the tank. Given the right conditions, cichlids can be some of the most colorful and most active fish in the tank. Some cichlids keep their bright colors only if they eat foods that contain certain orange, red, and yellow pigments, while others may lose their bright colors when exposed to conditions that are less than ideal.

Loaches

Loaches are excellent scavengers. They scour the bottom of the tank and eat food that has collected there. Native to Asia, they have long bodies and require little special care. Some popular species include skunk loaches, clown loaches, and blue loaches. Loaches prefer living with 2 or 3 of their own species.

Guppies

Guppies have beautiful fanlike tails that are speckled with a wide variety of colors and patterns. Most species of guppy are easy to raise. A few, however, require higher salt levels in water than most other fish can withstand. Guppies are unusual because they hatch from eggs kept inside the mother's body.

Angelfish

There are many species of marine angelfish but only one species of freshwater angelfish. This species is very hardy and attractive in an aquarium, but it can be aggressive with other fish.

Mollies

Two common varieties of mollies are the black molly and the sail-fin molly. Both are very hardy and beautiful fish to keep in an aquarium.

Plecostomus

The plecostomus is a scavenger fish. It eats algae on the sides of the aquarium and uneaten food that settles to the bottom of the aquarium. Take care when choosing a plecostomus for your aquarium; many varieties quickly outgrow their tanks. The clown and bristlenose plecostomus are good varieties for small aquariums.

Exploration 1 CD-ROM articles, continued

Corydoras Catfish

Like the plecostomus, catfish are hardy and easy to raise scavenger fish. Corydoras catfish scour the bottom of the tank for food. Special sinking foods may be required to keep them well fed. Frozen worms or sinking pellets are a good food source for these catfish.

Controlled Experiments

Scientific Method

The scientific method is a systematic way of asking questions, performing experiments, gathering data, drawing conclusions, and communicating results. A scientist begins by asking an investigative question, such as "How do birds know when it's time to migrate?" or "Do heavier objects fall faster than light ones?" Then he or she collects information or data about the question to form a hypothesis. The hypothesis is a possible explanation for an event. A good hypothesis is a statement or explanation that can be tested. The scientist then designs an experiment to test the hypothesis. As the experiment takes place, observations are recorded. By analyzing these observations, the scientist can draw conclusions about the hypothesis and communicate the results.

What if you wanted to find out how much fertilizer is best for growing a potted plant? One way to do this scientifically is to set up a controlled experiment. A controlled experiment is one in which only one factor or variable is changed at a time.

In this case, the only variable that changes is the amount of fertilizer each plant receives. Everything else—including the type and size of plant, amount of water, amount and intensity of sunlight, and type of soil—must stay the same. The experimental plants would be exactly like the control plant, except they would be given different amounts of fertilizer. This way, you can easily test the effect of fertilizer on the growth of the plant.

Recording Results

Scientists must carefully record the results of their work. You may decide to record your results in your notepad using a form that looks something like this.

Title of the experiment:

Description of the problem or question:

Research about the problem or question:

Hypothesis:

Variables to be controlled:

Experimental variable:

Description of experiment performed:

Data and observations:

Do these observations support the hypothesis?

Additional questions or possible hypotheses:

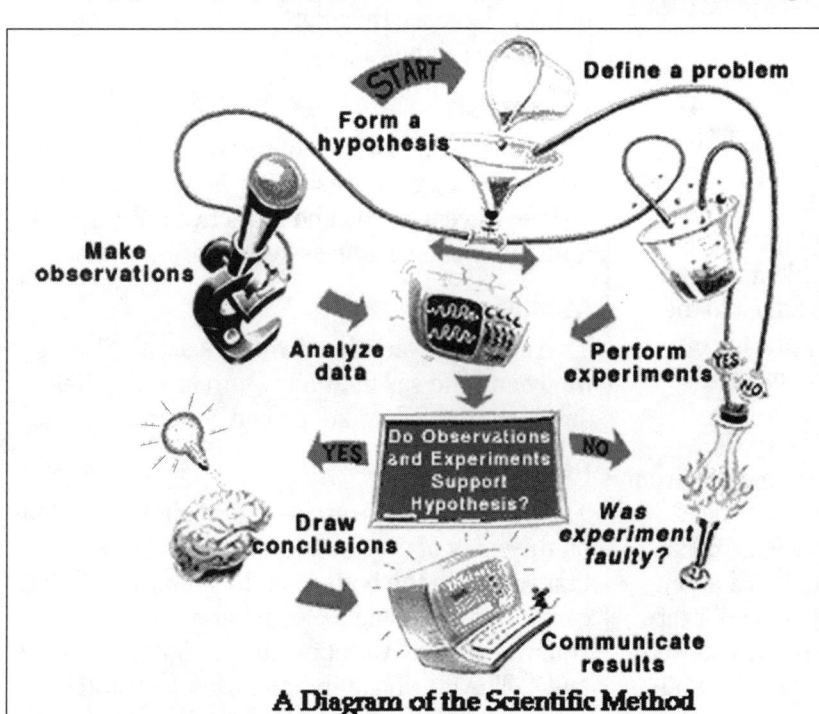

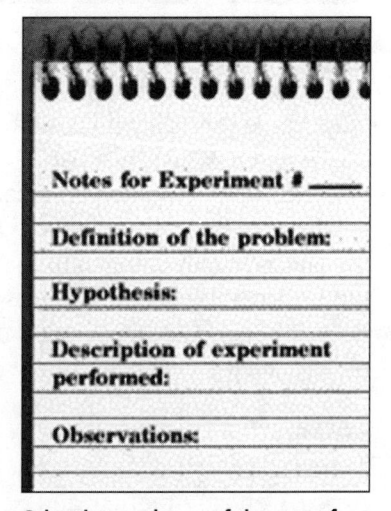

A Diagram of the Scientific Method

Scientists make careful notes of their observations and experimental results.

Exploration 2
Teacher's Notes

Shut Your Trap!

Key Concepts	Several variables may affect the health of living things. The Venus' flytrap gets nearly all the nutrients it needs from the insects it consumes; exposure to nutrient-rich soil will kill the plant.
Summary	Plant poachers have been taking Venus' flytraps from their native habitat in the bogs of North and South Carolina. The Bogs Are Beautiful Appreciation Society wants to prevent the Venus' flytrap from becoming extinct in the wild. The society needs to know the optimal growing conditions for the Venus' flytrap so that it can provide nursery owners with an easy and inexpensive method of growing the plant. Hopefully, the practice of poaching will then be reduced.
Mission	Determine the optimal growing conditions for the Venus' flytrap.
Solution	Several variables affect the growth of the Venus' flytrap. Nursery-cultivated flytraps thrive in conditions that most closely approximate the conditions of their natural habitat: 15 hours of light, 50 percent humidity, and no plant food. Adding plant food to the soil provides flytraps with an overabundance of nitrogen that kills the plants.
Background	The Venus' flytrap was once found in 21 counties surrounding Wilmington, North Carolina. Now, due to poaching and the destruction of the flytrap's natural habitat, the plant is found in only 11 of those counties. Even though flytraps can be grown in artificial environments, collecting the plants in the wild is cheaper, and poaching remains a problem. Because flytraps grow in sandy, boggy soil, they depend on periodic wildfires to clear patches in the bog for them to grow. However, the quick suppression of these fires by humans is contributing to the shortage of areas in which the flytraps can grow. Despite the work of government officials and the imposition of heavy fines for poaching, some scientists think that the flytraps may soon be found only in nature preserves and other protected areas.

Exploration 2 Teacher's Notes, continued

Teaching Strategies

This Exploration gives students the opportunity to conduct a controlled experiment in which several variables are combined to create optimal environmental conditions for the growth of a plant. Students must work efficiently because they have enough flies for only 10 experiments. Therefore, you may wish to encourage students to research the CD-ROM articles thoroughly before setting the variables in their control. Students who are having difficulty may require additional assistance. Make sure that they understand the purpose of a control. If necessary, review the diagram of the scientific method in the CD-ROM articles with students who are having difficulty.

As an extension of this Exploration, you may wish to have students discuss the consequences of allowing plant species to become extinct. Emphasize to students that finding an inexpensive method of growing the Venus' flytrap in an artificial environment could help to keep the plant from disappearing from the wild. Students may be interested to know that only 3 to 5 percent of the original habitat for carnivorous plants remains in the United States and that many species of carnivorous plants are already extinct. You might wish to have students contact an expert from a local college, university, or scientific institution to find out more about the preservation of disappearing plantlife.

Bibliography for Teachers

Albert, Victor A., Stephen E. Williams, and Mark William Chase. "Carnivorous Plants: Phylogeny and Structural Evolution," *Science,* 257 (5076): September 11, 1992, pp. 1491–1495.

Schnell, Donald E. *Carnivorous Plants of the United States and Canada.* Winston-Salem, NC: John F. Blair, Publisher, 1976.

Bibliography for Students

Doyle, Mycol. *Killer Plants: Venus' Flytrap, Strangler Fig, and Other Predatory Plants.* Los Angeles, CA: Lowell House, 1993.

Kite, L. Patricia. *Insect-Eating Plants.* Brookfield, CT: Millbrook Press, 1995.

Other Media

Death Trap
Videotape
Time-Life Video
777 Duke St.
Alexandria, VA 22314
703-838-7000
800-621-7026

In addition to the above videotape, students may find relevant information about the Venus' flytrap on the Internet. Interested students can search for articles with keywords such as *plants, carnivorous plants, insect-eating plants,* and *Venus' flytrap.*

Name _____ Date _____ Class _____

Exploration 2
Worksheet

Shut Your Trap!

1. What are Ms. Lily N. Lotus and the Bogs Are Beautiful Appreciation Society concerned about?

   _____
   _____
   _____
   _____

2. What are three variables that may be affecting the growth of the Venus' flytraps?

   a. _____

   b. _____

   c. _____

3. What will you use for a control in your investigations?

   _____
   _____

4. There are 24 possible variable settings for the experimental terrarium. However, you have only enough flies to conduct 10 experiments. What steps can you take to make sure that you find a solution before you run out of flies?

   _____
   _____
   _____

5. Form a hypothesis for how each variable affects the growth of the Venus' flytraps.

   Hypothesis 1: _____
   _____

   Hypothesis 2: _____
   _____

   Hypothesis 3: _____
   _____

EXPLORATION 2 • SHUT YOUR TRAP! 11

Name _____ Date _____ Class _____

Exploration 2 Worksheet, continued

6. Record your observations in the table below as you investigate each hypothesis.

Plant food	Humidity	Hours of light	Observations
Yes	0%	5	
		10	
		15	
		20	
	25%	5	
		10	
		15	
		20	
	50%	5	
		10	
		15	
		20	
No	0%	5	
		10	
		15	
		20	
	25%	5	
		10	
		15	
		20	
	50%	5	
		10	
		15	
		20	

Name _____ Date _____ Class _____

Exploration 2 Worksheet, continued

7. Were your experiments faulty in any way? If so, what steps did you take to correct them?

_____
_____
_____
_____
_____
_____
_____
_____
_____

Record your conclusions in the fax to Lily N. Lotus.

EXPLORATION 2 • SHUT YOUR TRAP! 13

Name _____ Date _____ Class _____

Exploration 2
Fax Form

FAX

To: Lily N. Lotus (FAX 910-555-5657)

From:

Date:

Subject: Optimal Growing Conditions for Venus' Flytraps

What is your recommendation? _____
_____
_____

✂·✂

For Internal Use Only

Please answer the following questions for my laboratory records. Scientists must always keep good records. *Dr. Crystal Labcoat*

During your experiments, which values proved to be optimal for the Venus' flytrap?

Optimal Values (select one per row) **EXPERIMENTAL VARIABLES**

5	10	15	20	HOURS OF LIGHT PER DAY
0	25	50		PERCENT HUMIDITY
YES	NO			ADD PLANT FOOD?

What effect did the hours of light per day have on the plants? Why?
_____
_____
_____

What effect did the percent humidity have on the plants? Why?
_____
_____
_____

SCIENCEPLUS INTERACTIVE EXPLORATIONS TEACHER'S GUIDE • LEVEL GREEN

Name _____ Date _____ Class _____

Exploration 2 Fax Form, continued

What effect did the plant food have on the plants? Why?

_____
_____
_____
_____
_____
_____
_____
_____
_____
_____

Shut Your Trap!

The following articles can also be found by accessing the computer graphic of the CD-ROM for Exploration 2:

- *Venus' Flytrap*
- *Controlled Experiments*

Exploration 2
CD-ROM Articles

Venus' Flytrap

What a Peculiar Plant!

Close to the ground in a North Carolina bog, a predator lies in wait. Only 15 cm high, the Venus' flytrap lures its prey—ants, flies, and other small insects—into a deadly trap. The Venus' flytrap spreads its leaves open, and when an unsuspecting insect touches hairs on the leaf, the leaf snaps shut. The more the insect struggles, the tighter the trap becomes! Once the trap has closed, the plant releases digestive enzymes that slowly dissolve the prey.

A fully grown Venus' flytrap may have 6–15 traps that are about 2.5 cm long.

How does the trap work? Each leaf has two sides that are lined with sensitive trigger hairs. If an insect touches two of these hairs, or touches the same hair twice in a short period of time, the two sides snap shut. A change in water pressure causes the leaves to close together. When the trigger hairs are touched, water moves from cells on the inside walls of the leaves to cells on the outside walls of the leaves. The inside of the trap becomes limp and causes the leaf to close.

At Home in the Bogs

The natural habitat of the Venus' flytrap is the sandy, boggy soil that lies between the pine woodlands and the dense bogs of North and South Carolina. It has adapted itself to the unique environment of this region. This environment provides the Venus' flytrap with plenty of water, sunlight, and air.

If you want to raise a Venus' flytrap on your own, you must carefully regulate its environment. You can grow a Venus' flytrap in a sphagnum-moss mixture or in a large tray of water. The roots need a lot of room, so make sure your young plant has space to grow. Always use plenty of distilled water or unpolluted rainwater; tap water will quickly kill a Venus' flytrap. The plant thrives in high humidity, so you should try to maintain a consistent humidity level above 30 percent. In addition, the Venus' flytrap is adapted to a warm and sunny environment. Your plant will need strong light and a temperature that stays between 21°C and 38°C.

Why Eat Bugs?

The Venus' flytrap is a green plant. That means it makes its food through photosynthesis like other plants. So why does it trap and digest insects? The answer lies in the nutrient-poor soil. The wetlands soil lacks essential nutrients, especially nitrogen, that the plant obtains from the insects it digests. Like most other carnivorous plants, the Venus' flytrap cannot survive in nutrient-rich soils; its insect-eating adaptations allow it to thrive only in nutrient-poor soil.

Reproduction: Survival of the Species

In ideal growing conditions, a Venus' flytrap can bloom and produce seeds in about 3 months. First a stalk grows out of the center of the plant. It can reach about 15 cm in height. The stalk then sprouts pink and white flowers that allow the Venus' flytrap to reproduce. When the plant begins to flower, its traps shrivel up. At the base of the flower, green seedpods form which are eventually released into the soil.

As the coastal plain of the Carolinas develops, the existence of the Venus' flytrap becomes more threatened. Poachers, cashing in on this popular plant, are seriously diminishing native populations. While the Venus' flytrap once grew wild in 21 counties, now the species survives in just 11 counties. Laws in North Carolina inflict stiff penalties on poachers, but still the species is in danger.

Controlled Experiments

Scientific Method

The scientific method is a systematic way of asking questions, performing experiments, gathering data, drawing conclusions, and communicating results. A scientist begins by asking an investigative question, such as "How do birds know when it's time to migrate?" or "Do heavier objects fall faster than light ones?" Then he or she collects information or data about the question to form a hypothesis. The hypothesis is a possible explanation for an event. A good hypothesis is a statement or explanation that can be tested. The scientist then designs an experiment to test the hypothesis. As the experiment takes place, observations are recorded. By analyzing these observations, the scientist can draw conclusions about the hypothesis and communicate the results.

What if you wanted to find out how much fertilizer is best for growing a potted plant? One way to do this scientifically is to set up a controlled experiment. A controlled experiment is one in which only one factor or variable is changed at a time.

In this case, the only variable that changes is the amount of fertilizer each plant receives. Everything else—including the type and size of plant, amount of water, amount and intensity of sunlight, and type of soil—must stay the same. The experimental plants would be exactly like the control plant, except they would be given different amounts of fertilizer. This way, you can easily test the effect of fertilizer on the growth of the plant.

Recording Results

Scientists must carefully record the results of their work. You may decide to record your results in your notepad using a form that looks something like this.

Title of the experiment:

Description of the problem or question:

Research about the problem or question:

Hypothesis:

Variables to be controlled:

Experimental variable:

Description of experiment performed:

Data and observations:

Do these observations support the hypothesis?

Additional questions or possible hypotheses:

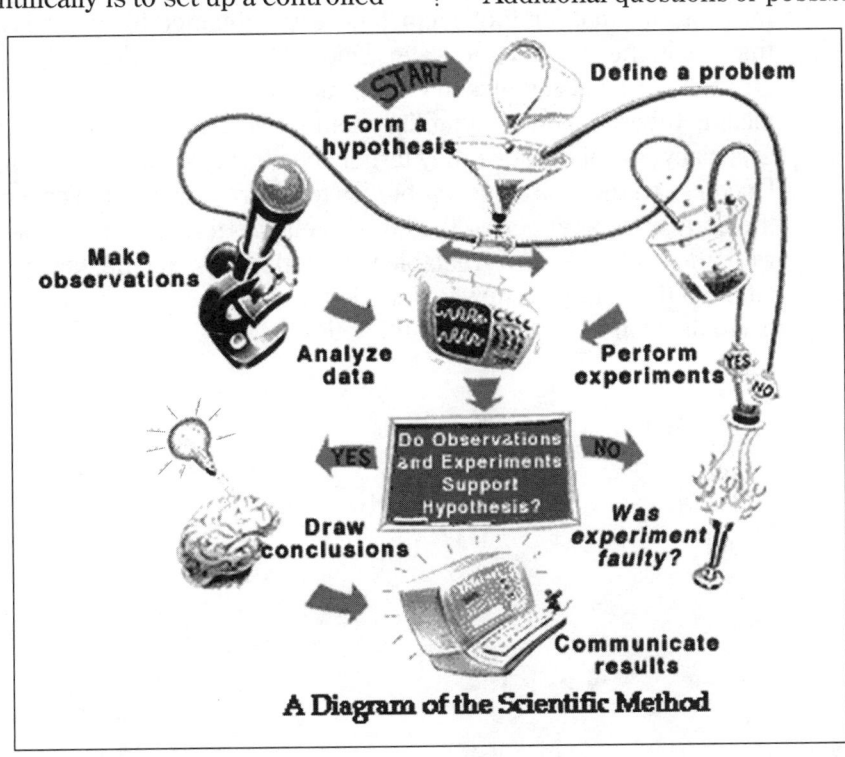

A Diagram of the Scientific Method

EXPLORATION 2 • SHUT YOUR TRAP! 17

Exploration 3
Teacher's Notes

Scope It Out!

Key Concepts	The spores of microorganisms can remain dormant for millions of years. Microorganisms can have helpful roles in the lives of other living creatures.
Summary	Dr. Viola Russ and her associates have been conducting experiments on amber. They have successfully extracted spores from a microorganism in a bee that was entombed in amber 25 million years ago. Dr. Russ now needs help to identify the microorganisms and their function and to discover whether they can be used to make antibiotics that will help fight modern diseases.
Mission	Identify the ancient microorganisms and determine their likely role in the life of the ancient bee.
Solution	The ancient microorganisms are rod-shaped bacteria of the kingdom Monera. They probably represent an ancient form of *Bacillus sphaericus*, a type of rod-shaped bacteria that aids bees in digestion and in fighting disease.
Background	Some students may ask why Dr. Russ is studying an ancient microorganism for use as an antibiotic. Explain that antibiotics are substances produced by microorganisms. Antibiotic substances are obtained from bacteria and fungi that live in the air, soil, and water. Doctors use antibiotics to fight various diseases that are caused by some harmful microorganisms. These diseases include tuberculosis, meningitis, pneumonia, and scarlet fever. Antibiotics are especially useful because they target specific cells without damaging others. Doctors can therefore use antibiotics to destroy harmful microorganisms in the body without affecting the body's cells. As a result, many diseases that were once fatal are now treatable with antibiotics. Scientists like Dr. Russ are interested in studying newly identified microorganisms because of the potential to develop new life-saving antibiotics.

Exploration 3 Teacher's Notes, continued

Teaching Strategies

Successful completion of this Exploration depends heavily on students researching the CD-ROM articles thoroughly. You may wish to encourage students to make use of the Notepad function as they conduct their research. In addition, student groups may benefit from a blackline master version of the articles. Those articles are available on pages 24–26 of this guide. Recommending that students reread the CD-ROM articles once the ancient microorganism has been identified may be especially helpful.

As an extension of this Exploration, you may wish to have students find out more about how to limit the harmful effects of microorganisms. For example, interested students could research the processes of pasteurization and sterilization and report their findings to the class in the form of an oral report. Students might also wish to contact a local restaurant to find out what steps are taken by the kitchen staff to prevent harmful microorganisms from spoiling food.

Bibliography for Teachers

Bodanis, David. *The Secret Garden: Dawn to Dusk in the Astonishing Hidden World of the Garden.* New York City, NY: Simon and Schuster, 1992.

Sagan, Dorion, and Lynn Margulis. *Garden of Microbial Delights: A Practical Guide to the Subvisible World.* Boston, MA: Harcourt Brace Jovanovich, Publishers, 1988.

Bibliography for Students

Dixon, Bernard. *Power Unseen: How Microbes Rule the World.* New York City, NY: W.H. Freeman & Company Limited, 1994.

Lovett, Sarah. *Extremely Weird Micro Monsters.* Santa Fe, NM: John Muir Publications, 1993.

Other Media

Organizing Protists and Fungi
Software (Apple II, MS-DOS, or Macintosh)
Queue, Inc.
338 Commerce Dr.
Fairfield, CT 06432
203-335-0906
800-232-2224

Simple Organisms: Bacteria (rev.)
Film and videotape
Coronet/MTI
P.O. Box 2649
Columbus, OH 43216
800-777-8100

Interested students may find relevant information about microorganisms and their roles on the Internet. Suggest that students search the Net with keywords such as the following: *microorganisms*, *antibiotics*, *bacteria*, and *fungus*.

EXPLORATION 3 • SCOPE IT OUT! 19

Name _____ Date _____ Class _____

Exploration 3
Worksheet

Scope It Out!

1. What does Dr. Viola Russ need to know about the ancient microorganisms?
   _____
   _____
   _____
   _____

2. What does Dr. Russ intend to do with the results?
   _____
   _____
   _____
   _____

3. What will you use to conduct your investigation?
   _____
   _____
   _____
   _____
   _____
   _____

4. What do the ancient microorganisms look like under the microscope?
   _____
   _____
   _____
   _____
   _____
   _____

Name _____ Date _____ Class _____

Exploration 3 Worksheet, continued

5. Use the table below to record your observations of each slide of microorganisms. Make sure that you write out the name of each microorganism in the left-hand column.

Protista	Observations
1.	
2.	
3.	
4.	

Monera	Observations
5.	
6.	
7.	

Fungi	Observations
8.	
9.	
10.	

EXPLORATION 3 • SCOPE IT OUT! 21

Name _____ Date _____ Class _____

Exploration 3 Worksheet, continued

6. Which modern microorganisms look the most like the ancient microorganisms?
   _____
   _____
   _____

7. How might you classify the ancient microorganisms and find out more about their likely role in the life of the ancient bee?
   _____
   _____
   _____
   _____
   _____
   _____
   _____
   _____
   _____
   _____

Record your conclusions in the fax to Dr. Russ.

Name _____ Date _____ Class _____

Exploration 3
Fax Form

FAX

To: Dr. Viola Russ (FAX 805-555-2266)

From:

Date:

Subject: Ancient microorganism classification and probable function

What are the ancient microorganism's classification and function? _____
_____
_____
_____

For Internal Use Only

Please answer the following questions for my laboratory records. Scientists must always keep good records. *Dr. Crystal Labcoat*

Which of the following may be used to classify the ancient microorganisms? Place an X in the left-hand column beside the correct answer(s).

KINGDOM	PROTISTA	MONERA	FUNGI
	Euglena	Round-shaped bacteria	Mildew
	Paramecium	Rod-shaped bacteria	Mold
	Amoeba	Spiral-shaped bacteria	Yeast
	Algae		

What role did this microorganism most likely play in the life of the ancient bee?
_____
_____
_____

EXPLORATION 3 • SCOPE IT OUT! 23

Scope It Out!

The following articles can also be found by accessing the computer graphic of the CD-ROM for Exploration 3:

- *The World of Microorganisms*

Exploration 3
CD-ROM Articles

The World of Microorganisms

What Is a Microorganism?

Microorganisms are so small that they can be seen only with a microscope. Microorganisms can be found in almost every environment on Earth. They can be found in air, soil, water, and other living things. Many microorganisms form the basis of food chains. Some cause diseases, such as malaria, AIDS, and strep throat. Other microorganisms help larger organisms to survive by aiding in digestion and other life processes.

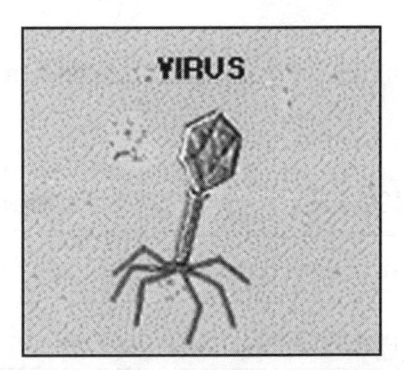

Viruses

Among the smallest of microorganisms, viruses can be seen only with a powerful electron microscope. Because viruses cannot reproduce without infecting living cells, cannot make proteins, cannot use energy, and are not made of cells, biologists do not consider viruses to be living organisms. However, viruses can harm living organisms and in fact are responsible for a number of diseases. Viruses are made up of strands of genetic material inside a protein coat. The information from a cell's genetic material is responsible for all of the cell's life processes.

When a virus infects a living cell, it replaces the cell's genetic information with its own, which causes the cell to stop functioning properly.

Viruses cause many common diseases, such as the cold, influenza (the flu), and chickenpox. Generally, the human body can defend itself against these infections. Some viral diseases can be prevented by immunizations, which are usually injections of weak or dead viruses that "trick" the body into behaving as if it has already had the disease. That way, the body may be able to quickly fight off a dangerous, invading virus in the future. Other viral diseases are more deadly. HIV is the virus that causes AIDS. HIV attacks the body's immune system, making it difficult for the body to defend itself against infection. Researchers continue to search for a vaccination against HIV and a cure for AIDS.

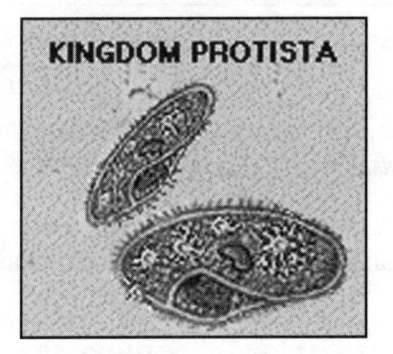

Kingdom Protista

Most organisms that belong to the kingdom Protista are single-celled. These organisms have some of the characteristics of plants and some of the characteristics of animals. They all live in moist environments, and they display a wide variety of shapes and structures. Consider the protists described below.

Amoeba—Amoebas are found in fresh water, salt water, and soil. They move by stretching out pseudopodia of intercellular liquid. Pseudopodia bulge out from the rest of the amoeba and can engulf food particles for the amoeba to digest. Amoebas reproduce by dividing in two. Most live freely in water or soil, while some, such as the amoeba that causes dysentery, infect other organisms.

Dinoflagellate—This microorganism is found in the ocean. It moves by using two flagella that cause it to spin through the water like a top. Some produce poisons that can harm other marine organisms, but most dinoflagellates are an abundant source of food for other species. A few of these kinds of organisms give off a glow much like fireflies!

Euglena—The single-celled Euglena can make its own food like a plant or absorb nutrients from its environment. The Euglena lives in fresh water and moves by wiggling its tail, which is called a flagellum. The Euglena has an eyespot that helps it orient its movements toward light.

Green algae—Most green algae live in fresh water and are single-celled; some, however, are multicellular. Green algae contain the same pigments that are found in most plants. For this reason, many scientists believe that green algae are the ancestors of the plant kingdom.

Paramecium—The Paramecium lives in fresh water and uses thousands of fine hairs, called cilia, to move and feed. The cilia beat in waves that cause the Paramecium to spin as it moves through the water. The cilia also sweep food into a narrow groove where the food is engulfed by a small, liquid-filled bubble called a vacuole. As the vacuole moves through the Paramecium, the food inside it is digested.

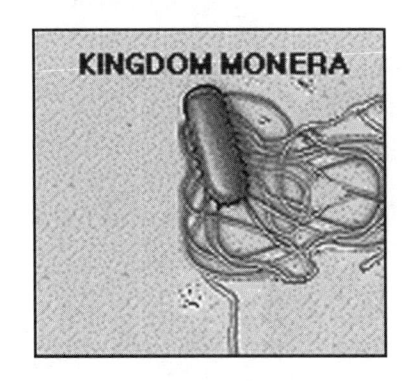

Kingdom Monera

The most common monerans are bacteria, which are single-celled microorganisms that do not have a nucleus but do have a cell wall. They are the oldest, simplest, and most abundant form of life on Earth. Bacteria can live in a variety of conditions, from hot springs to icy mountaintops. Some bacteria form spores in which most life functions stop until environmental conditions are able to support the bacteria again. Bacteria are classified according to their shape. Cocci are round, bacilli are rod-shaped, and spirilla are shaped like spirals. Bacteria reproduce quickly by dividing into two cells. Under proper conditions, one bacterium can divide every 20 minutes. Some forms of bacteria cause disease either by destroying living cells or producing dangerous toxins. Strep throat is a common bacterial disease caused by streptococcus bacteria.

Other bacteria perform useful functions, such as breaking down dead matter and other wastes. Bacteria are even used in producing foods such as cheese and yogurt.

Kingdom Fungi

Fungi are microorganisms that break down once-living material and absorb food through their cell walls. The bodies of fungi are made of a network of threadlike structures called hyphae. Reproductive spores are attached to the ends of the hyphae. Once released, these spores can survive for long periods of time before they find the proper environmental conditions and begin to divide. Most fungi need a moist environment to survive, and fungi frequently appear after a rainfall or in damp areas. Common kinds of fungi include mold, mildew, yeast, puffballs, and mushrooms. Mold appears as a fuzzy substance on the surface of once-living material, such as fruit or other foods. Mildew appears as a powdery or downy covering and often grows on leaves. Yeasts are a kind of fungi that form sacs of spores. Bakers use a kind of yeast to make bread rise. Puffballs and mushrooms form club-like structures in which the cap contains the spores.

Helpful or Harmful?

Are microorganisms helpful or harmful? The answer depends on the microorganism in question. Some microorganisms cause deadly diseases. The bubonic plague, typhoid, and cholera are caused by bacteria. AIDS, polio, measles, and mumps are caused by viruses. One form of dysentery is caused by a kind of amoeba.

Other microorganisms are helpful. *Penicillium* is a kind of fungi that fights bacterial disease. Yeast, another type of fungi, is used in baking and brewing. A certain bacteria, *Escherichia coli*, lives in the digestive tracts of humans and aids in the breakdown and digestion of food. Another bacteria, *Bacillus sphaericus*, plays a similar role in the digestive tracts of bees. Several forms of bacteria are used to make products such as cheese and yogurt.

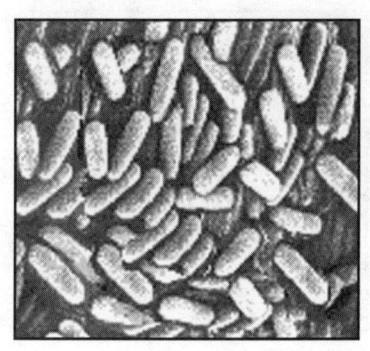

Escherichia coli

Most microorganisms cannot be classified as helpful or harmful. These organisms are simply a part of the environment, carrying out roles that we have yet to learn much about.

Microorganisms Trapped in Time

In 1991, Dr. Raul Cano announced that he had revived microorganisms that were between 25 and 40 million years old. These microorganisms were trapped in the gut of a bee that had been preserved in amber. The amber formed when resin flowed from a tree and hardened. The bee died quickly in the amber, but the microorganisms inside its body did not. Instead, some of the microorganisms became dormant.

Dr. Cano says that he revived the dormant microorganisms by exposing them to the proper nutrient-rich environment. Critics of Dr. Cano's work do not believe that such ancient life can be revived. They say that modern microorganisms must have contaminated Dr. Cano's results. Dr. Cano contends that he followed strict guidelines to prevent any contamination. In addition to the microorganisms found in the ancient bee, Dr. Cano claims to have revived over 2000 other kinds of microorganisms that were trapped in a variety of insects.

What's the Matter?

Key Concepts Each element has its own unique melting point and its own unique boiling point. Experimental procedures should be planned in advance to avoid unproductive investigations.

Summary Several researchers have been investigating the physical properties of molten lava, but, unfortunately, the tip of their temperature-measuring device, the lava analyzer, has melted. They need to know what type of metal will make the most practical replacement for the tip of the lava analyzer.

Mission Determine which metal will make the best tip for the lava analyzer.

Solution The high melting point of titanium, its availability, and its relatively affordable price make it the best metal to use for the tip of the lava analyzer. Tungsten and platinum are reasonable choices because both can withstand the heat of the molten lava. However, their scarcity and costliness make these metals less practical than titanium.

Background Students may be interested to know that scientists are actually studying the Kilauea volcano in Hawaii. The Hawaiian Volcano Observatory (HVO) is located on the summit of Kilauea, one of the most active volcanoes in the world. Frequent eruptions at Kilauea and nearby Mauna Loa make the HVO an ideal location for studying volcanoes. As a result, many of the world's most famous vulcanologists have studied at this observatory. In addition, many important new techniques and instruments for monitoring volcanoes have originated at the HVO.

In addition to the HVO, there are two other major sites in the United States for studying volcanoes, the Cascades Volcano Observatory and the Alaska Volcano Observatory. At all three of these sites, scientists monitor geophysical changes involving seismicity, ground movements, gas chemistry, hydrologic conditions, and the activity between and during eruptions. In addition, the scientists carefully analyze data to warn of possible eruptions and of the specific hazards related to those eruptions. When eruptions do occur, these scientists are fully prepared to study the eruptive behavior, identify the activities that led to the eruption, and define the processes by which its deposits are left behind.

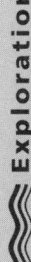

Exploration 4 Teacher's Notes, continued

Teaching Strategies

One of the purposes of this Exploration is to allow students to make decisions about how to conduct an investigation in the most efficient manner possible. One activity, the determination of the mass and volume of each metal, is a distracter. With careful forethought, students may be able to predict ahead of time that such an activity would have no relevance to the task at hand. Encourage students to establish in advance a procedure for solving Dr. Stokes's problem. Be sure that students include a careful study of the CD-ROM articles in their procedure.

As an extension, you may wish to encourage students to do further research on one of the metals described in the Exploration. Students could focus their research on how the metal is used in its solid, liquid, and gaseous states in important everyday products. Suggest that students report their findings to the class in the form of a poster or other visual display.

Bibliography for Teachers

Barber, Jaqueline. *Solids, Liquids, and Gases: A School Assembly Program Presenter's Guide.* GEMS Series, edited by Lincoln Bergman and Kay Fairwell. Berkeley, CA: Lawrence Hall of Science, 1986.

Davies, Paul, and John Gribbin. *The Matter Myth.* New York City, NY: Simon & Schuster/Touchstone, 1992.

Bibliography for Students

Berger, Melvin. *Solids, Liquids and Gases: From Superconductors to the Ozone Layer.* New York City, NY: G.P. Putnam's Sons, 1989.

Cooper, Christopher. *Matter.* Eyewitness Science series. New York City, NY: Dorling Kindersley, Inc., 1992.

Darling, David. *From Glasses to Gases: The Science of Matter.* New York City, NY: Dillon Press, 1992.

Other Media

Particles in Motion: States of Matter
Filmstrip
National Geographic Society Educational Services
P.O. Box 98019
Washington, DC 20090-8019
800-368-2728

Interested students may also find relevant information about states of matter on the Internet using keywords such as the following: *physical changes*; *matter*; and *solids, liquids, and gases*. Students may also be interested in Internet resources about *lava, volcanoes, vulcanologists,* and *volcano observatories*.

Name _____ Date _____ Class _____

Exploration 4
Worksheet

What's the Matter?

1. What problem does Dr. Stokes need you to help him solve?

   _____
   _____
   _____

2. Dr. Labcoat has gathered a set of metal samples for you to analyze. List the name of each metal and record its melting point and boiling point in the following data chart:

Metal	Name of metal	Melting point (°C)	Boiling point (°C)
Cu			
Sn			
Pt			
Ti			
W			
Al			

3. How might the information in the table be useful to you in solving Dr. Stokes's problem?

   _____
   _____
   _____
   _____

4. What additional information could help you solve the problem?

   _____
   _____
   _____
   _____

EXPLORATION 4 • WHAT'S THE MATTER?   29

Name _____ Date _____ Class _____

Exploration 4 Worksheet, continued

5. How might you find this information?

   _____
   _____
   _____
   _____

6. Record helpful data here, continuing on the back of the page if necessary.

   _____
   _____
   _____
   _____
   _____
   _____
   _____
   _____
   _____
   _____

7. When you have finished, evaluate the procedure that you used to complete this Exploration. What would change about your procedure? Did you perform any activities that were not useful to you? If so, which ones?

   _____
   _____
   _____
   _____

8. How could you improve your procedure?

   _____
   _____

Record your conclusions in the fax to Dr. Stokes.

SCIENCEPLUS INTERACTIVE EXPLORATIONS TEACHER'S GUIDE • LEVEL GREEN

Name _____ Date _____ Class _____

Exploration 4
Fax Form

FAX

To: John Stokes, Ph.D. (FAX 080-555-9822)

From:

Date:

Subject: Metal recommendation

What is your recommendation? _____
_____
_____
_____

✂·

For Internal Use Only

Please answer the following questions for my laboratory records. Scientists must always keep good records. Dr. Crystal Labcoat

Please indicate your metal selection here: _____

How do the particles of this metal behave during the following phases:

solid? _____
_____
_____

liquid? _____
_____
_____

gas? _____
_____
_____

EXPLORATION 4 • WHAT'S THE MATTER?   31

What's the Matter?

The following articles can also be found by accessing the computer graphic of the CD-ROM for Exploration 4:

- *Matter*
- *Metals*
- *Volcanoes*

Exploration 4
CD-ROM Articles

Three states of matter: solid, gas, and liquid

Matter

The Basics of Matter

Matter is defined as anything that takes up space and has mass. Matter is made up of small particles called atoms and molecules. Different atoms and molecules have different physical and chemical properties.

Odor, color, taste, hardness, mass, density, melting point, and boiling point are just a few examples of *physical properties*. All of these properties can be observed and even measured without changing the composition of the matter. For example, you can observe the melting point of an ice cube without the ice cube changing into a different kind of matter. This is not the case with chemical properties.

Chemical properties can be observed only when one kind of matter changes into another kind of matter. For example, when you add baking soda to vinegar, a different kind of matter is formed—carbon dioxide gas. A chemical property of baking soda, then, is the formation of a gas when vinegar is added.

States of Matter

Most matter you will encounter exists in one of three states: solid, liquid, or gas.

Solids are rigid, cannot be noticeably compressed, and usually have distinct boundaries. The atoms and molecules in solids are tightly arranged, like people sitting in rows in a theater.

Liquids flow, cannot be noticeably compressed, and usually have a boundary with air. In liquids, atoms and molecules move past each other but remain within a boundary. A liquid can be compared to people moving around in a theater's lobby.

Gases flow, can be compressed, and have no boundary with air. Atoms and molecules in gases spread apart, much like people leaving the theater and going in different directions.

Changing States of Matter

If you leave a candle sitting in the hot sun, it will change from a rigid stick to an oozing liquid. This transformation is called a *change of state* or a *phase change*. The candle changes state but does not change its composition. The oozing liquid is still wax. This means that the basic particles that make up the wax do not change but the form of the wax does. For another example, consider water. A water molecule remains a water molecule whether it is ice (solid), water (liquid), or vapor (gas). The water has simply undergone changes of state. And since these changes of state do not change the composition of the water, they are considered physical changes.

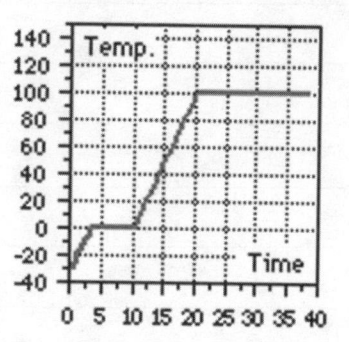

Graph showing the roles of temperature and time in changing states of matter

32   SCIENCEPLUS INTERACTIVE EXPLORATIONS TEACHER'S GUIDE • LEVEL GREEN

The state of a particular type of matter depends primarily on its temperature. When a solid is heated, its temperature rises until it reaches its melting point. The melting point is the temperature at which a solid changes to a liquid. The melting-point temperature remains constant until all of the solid has changed to a liquid. As the liquid is heated, the temperature begins to rise again until it reaches its boiling point. The boiling point is the temperature at which a liquid changes to gas. The temperature remains constant at the boiling point until all of the liquid has changed to gas.

The Nature of Matter

The kinetic molecular theory of matter states that matter is composed of very tiny particles that are in constant motion. This theory helps explain some of the physical properties of solid, liquid, and gaseous matter.

The particle model of matter assumes four basic principles about matter.

1. All matter is composed of tiny particles called atoms.
2. Each element is made up of the same kind of atoms, and the atoms of one element are different from the atoms of all other elements.
3. Atoms cannot be divided, created, or destroyed.
4. Atoms of elements combine in certain ratios to form compounds.

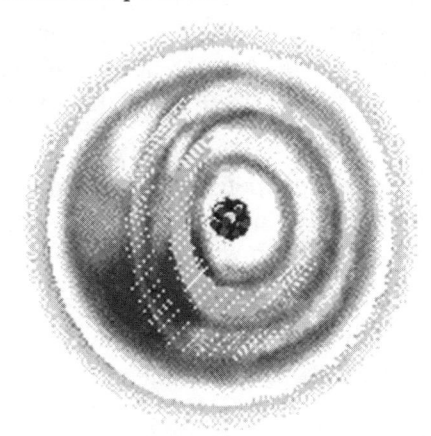

Model of an atom

Although modern research has shown that not all matter is composed of atoms and that atoms are composed of even smaller particles, this model is still a useful way to understand the nature of matter.

Periodic Table of Elements

The periodic table of elements is a chart that shows all of the known elements arranged by similar physical and chemical properties. Elements with similar properties are grouped together in vertical columns. Major classes of elements include the alkali metals, lanthanides, actinides, halogens, and noble gases. In addition to classifying elements by properties, the periodic table also helps scientists make predictions about the properties of new elements by comparing them to the properties of known elements already in the chart.

Each square on the chart describes an element by listing the element's atomic number, chemical symbol, element name, and atomic mass. The atomic number identifies the number of protons in an atom's nucleus. Atomic mass is the average mass of the element. The chemical symbol is an international abbreviation for the element. Sometimes the symbol is the first letter of the element's name, such as *C* for carbon. However, the symbol may also come from a Greek or Latin word for the element. The symbol for lead is Pb, which comes from *plumbum*, the Latin word for "lead." Some elements are named for a place or individual; for example, Einsteinium was named in honor of the physicist Albert Einstein.

Metals

Aluminum

Atomic Number: 13
Atomic Mass: 26.9815
Description: silvery white

Clays containing aluminum were used for making pottery in the Middle East over 7000 years ago. Other aluminum compounds were also used in ancient Egypt and Babylonia.

Aluminum is an extremely useful metal, primarily because it is lightweight. Despite this, aluminum mixtures can be made strong enough to replace many heavier metals. Aluminum is also long lasting and less expensive than many other metals. It conducts heat and electricity well and does not rust easily. Aluminum is an excellent reflector of heat, so it is useful in building insulation and roofing materials. Most aluminum produced today is made from bauxite ore, which makes up about 8 percent of the Earth's crust. Large deposits are found throughout the world.

It takes less energy to recycle used aluminum than to produce new aluminum. Recycling efforts provide over half the aluminum supply used in making new aluminum products, mostly aluminum cans. Aluminum is used in many other objects such as airplanes, automobile parts, household utensils, foil, and building supplies.

Copper

Atomic Number: 29
Atomic Mass: 63.546
Description: reddish

Copper is extracted from veins of copper ore in the Earth. Once refined, copper is quite malleable, which means it can be easily shaped or formed. People have used copper since prehistoric times to make tools, utensils, jewelry, and weapons. In modern times, copper is used in piping, cookware, wire, and coins. There are more than 160 known minerals that contain copper. One of these minerals is yellow chalcopyrite, otherwise known as fool's gold.

Copper is extremely useful because it readily conducts heat and electricity. Silver is the only metal that conducts electricity better than copper at room temperature. Copper also resists rust and is relatively inexpensive. It can be mixed with other metals to form strong alloys such as brass and bronze.

Gold

Atomic Number: 79
Atomic Mass: 196.967
Description: bright yellow, soft

Gold has been a symbol of wealth for thousands of years. It is scarce, but can be found worldwide. Scientists believe that geologic activity moved gold from the Earth's inner depths through cracks and fissures to the Earth's surface. This explains why gold is found in igneous rocks and sea water throughout the world. Deposits of gold are small and rarely pure. Gold is normally found mixed with other metals, such as silver, lead, or copper. Although gold is found in all sea water, its extraction from sea water is expensive.

Gold does not tarnish, rust, or corrode like most other metals. Gold is also the most malleable metal, and it can be hammered into sheets that are so thin they are transparent. Gold is often mixed with other metals, such as silver, copper, and nickel, to form alloys. The amount of gold in each of these alloys is measured in karats. Gold is most often used for decorative purposes, such as jewelry, but is also used in sophisticated electronics, such as those used on spacecraft and satellites.

Iridium

Atomic Number: 77
Atomic Mass: 192.22
Description: hard, brittle, silvery

Iridium is from the same family as platinum and is one of the hardest metals known. Ores and deposits containing pure iridium are extremely scarce, and separating it from other compounds is a very complicated and costly process. Iridium was named after the Greek goddess of the rainbow, Iris, because the salts of this metal are brightly colored. Generally, iridium is mixed with other metals to add hardness to the resulting alloy. Iridium is used to make jewelry, surgical instruments, pen points, and electrical contacts.

Iron

Atomic Number: 26
Atomic Mass: 55.847
Description: silvery white, lustrous, magnetizable

Iron composes about 5 percent of the Earth's crust, making it one of the most abundant elements. Since prehistoric times, iron has been extracted from its ores and shaped for a wide variety of uses. It is estimated that around 1200 B.C., people discovered how to melt iron and forge it into tools and weapons. This discovery began a period in history called the Iron Age.

Iron is still widely used today because it is plentiful and relatively inexpensive. Common items such as frying pans and pipes are made of iron. Iron can also be mixed with other metals to make very strong alloys, such as steel.

Another widely used iron alloy is cast iron, which is valuable because of its hardness, low cost, and ability to absorb shock. Cast iron is a compound that contains from 2 to 4 percent carbon and 1 to 3 percent silicon. Because of its high carbon content, cast iron can be shaped only by melting it and pouring the liquid metal into a mold. This casting process is used to make items such as automobile engines, fire hydrants, and construction materials.

Lead

Atomic Number: 82
Atomic Mass: 207.29
Description: silvery, soft

Lead is one of the heaviest metals known, but it is so soft that it can be scratched with a fingernail. It is easily molded into a wide variety of objects, such as pipes, sculptures, fishing weights, and bullets. Because lead is plentiful, it is an inexpensive metal to obtain.

Some lead compounds are poisonous, limiting its use in many products. Until the 1970s, lead was used as a white pigment in house paint. Many children became sick from eating the peeling paint chips that contained lead. Other substances, including titanium dioxide and zinc, are now used in paint instead of lead.

Platinum

Atomic Number: 78
Atomic Mass: 195.08
Description: silver-white

Platinum is found worldwide and is usually mixed with other metals. However, it is not an abundant metal. It makes up only about a millionth of 1 percent of the Earth's crust. Since deposits of platinum are so small, the price of this metal is similar to the high price of gold.

Platinum is easily shaped, does not tarnish, conducts electricity, and resists high temperatures and chemicals. These properties make it a useful metal for missile cones, jet-engine fuel nozzles, and dental fillings. In addition, platinum is often used for making jewelry. When combined with gold, it makes an alloy known as "white gold."

Silver

Atomic Number: 47
Atomic Mass: 107.868
Description: brilliant white

Silver has long been popular because of its beauty and usefulness. It can be found in veins of pure silver or, more commonly, with other elements. Although silver is considered a rare metal, it costs less than platinum or iridium.

For thousands of years, silver has been used as money. A soft, malleable metal, it is usually combined with other metals to make tableware, coins, and jewelry. High-grade silver jewelry is usually made of sterling silver, which is an alloy of silver and copper.

One of silver's most valuable properties is that it conducts heat and electricity better than any other known metal. For this reason, it is often used to make electrical contacts. Other industrial uses for silver include photographic film, batteries, and coatings for the backs of mirrors. Silver is also used to plate, or coat, less expensive metals.

Tin

Atomic Number: 50
Atomic Mass: 118.69
Description: silvery white

Many people associate tin with tin cans. But a tin can is actually a steel can coated with tin. Because tin is nontoxic, malleable, and resists corrosion, it makes an excellent protective coating for other metals. Tin is also used to make alloys such as bronze (tin and copper) and pewter (tin, copper, and antimony).

Tin is usually extracted from its main ore, cassiterite. Because tin occurs less frequently in nature than copper and aluminum, it is more expensive than these metals.

Titanium

Atomic Number: 22
Atomic Mass: 47.88
Description: strong, silvery gray

Titanium is the ninth most abundant element on Earth. Rock samples from the moon show that this metal is present there, too. Titanium is as strong as steel and 45 percent lighter. It resists corrosion and high temperatures, which makes it especially useful for airplanes and spacecraft. Most white paint now contains highly reflective titanium dioxide rather than more toxic lead compounds. Surgical-implant devices are also made of titanium and its alloys because they are compatible with human tissue.

Tungsten
Atomic Number: 74
Atomic Mass: 183.84
Description: hard, brittle, steel gray to white

Tungsten is the twenty-third most abundant element on Earth, which makes it less common than iron, titanium, or zinc. It has the highest melting point of any metal, making it ideal for high-temperature uses such as light-bulb filaments, fireproof cloth, and heating elements for electric furnaces. In its pure state, tungsten is very hard and brittle. Because of these qualities, it is often used in alloys to add hardness and strength to softer metals. However, tungsten is always found in an impure state in nature and is costly and difficult to process and refine.

Tungsten filament

Zinc
Atomic Number: 30
Atomic Mass: 65.39
Description: bluish white

Zinc ranks fourth, after steel, aluminum, and copper, among the metals most commonly produced. It replaces lead in paint because it is nontoxic, reflects light, and resists mold and fungus. Zinc is also mixed in small quantities with copper and other metals to make bronze and brass. Zinc is fairly inexpensive and is widely distributed throughout the world.

Zinc can be used as a protective coating that keeps steel from rusting. In a process called *hot-dip galvanizing*, the steel is heated, dipped into a vat of molten zinc, and then passed through a cooling tower, where the zinc coating hardens. Common items made of galvanized steel include heating ducts, storage tanks, light poles, fencing, and highway guardrails.

Zirconium
Atomic Number: 40
Atomic Mass: 91.22
Description: grayish white

Zirconium is a rare, soft metal that comes from the mineral zircon, which often looks very much like diamonds. Zirconium is found only in very small quantities on Earth. Traces of it have also been identified in the sun, other stars, and some meteorites.

Because zirconium has a very high melting point and resists corrosion and the absorption of atomic particles, it is widely used in the field of nuclear energy. It is also used in rocket engines, high temperature furnaces, and medical hardware such as pins and screws used to repair broken bones.

Nuclear energy symbol

Volcanoes

Eruptions on Earth

A volcano is an opening in the Earth's surface through which gases and hot molten rock erupt. Molten rock, called magma, is pushed up through the Earth by explosive gases deep within the Earth's mantle. Sometimes when the magma reaches the Earth's surface, it is called lava.

Vulcanologists are scientists who study volcanoes. Vulcanologists estimate that there are 40,000 volcanoes on Earth, three-fourths of which are located under water. Volcanoes do not occur only on Earth, however. Scientists have discovered volcanoes on Mars and Venus, as well as on Io and Europa, two of Jupiter's moons.

Types of Volcanoes

Scientists classify volcanoes based on how the eruptions form mountains. In some eruptions, lava simply pours from an opening in the crust. Hardened lava gradually builds up, forming a mountain, or cone, with gently sloping sides. This type of volcano is called a *shield volcano*. Mona Kea, in Hawaii, is a shield volcano. Despite its gentle slope, it rises 10,203 m from the ocean floor to its summit, forming the largest mountain on Earth. Measured from its base to the ocean floor, it is almost 1400 m taller than Mount Everest.

Not all eruptions are as gentle as those that form shield volcanoes. When magma is thick, pressure from trapped gases within the magma may build up, causing an explosive eruption to take place. Such an explosion can spew lava and tephra—volcanic rock, cinders, and ash—several kilometers into the air. This eruption creates a *cinder-cone volcano,* which has steep sides and a narrow base. Once the pressure is released, this type of volcano usually becomes inactive. One example of a cinder-cone volcano is Paricutín, in Mexico. This volcano grew to 100 m in its first five days of eruption in 1943.

Some volcanoes, called *composite volcanoes,* are formed from alternating layers of lava and tephra. Many well-known volcanoes, such as Vesuvius and Mount St. Helens, are composite volcanoes. They usually have steep slopes and broad bases. Because of their layered construction, composite volcanoes are sometimes referred to as *stratovolcanoes.*

The eruption of a composite volcano

Lava

Lava varies in color from dark red to light yellow, depending on its chemical composition and temperature. The temperature of lava ranges from 600°C to 1200°C. Lava keeps heat so well that it may take up to a month for a 1 m thick flow to cool. As it cools, lava hardens into many different forms. Smooth, folded sheets of lava are called pahoehoe. Jagged, rough sheets are called aa. Lava also cools into domes, steep hills, and tunnels.

The Ring of Fire

Volcanoes are often located where the large rigid plates of Earth's crust meet. Many of these plates are located along the ocean floor. So many volcanoes encircle the Pacific Ocean plate that this region is called the Ring of Fire.

One of the volcanoes in this region is Kilauea (kil uh WAY uh), on the island of Hawaii. Kilauea lies far from a plate boundary, but at a place where a huge column of magma rises from within the Earth. Kilauea, which means "rising smoke cloud," has a deep, round crater at its center. When the volcano erupts, the crater fills with a lake of boiling lava.

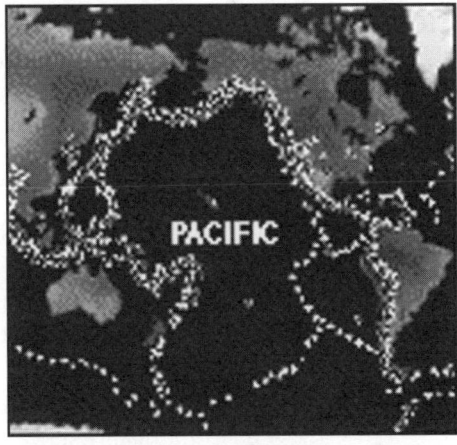

The Ring of Fire is a line of volcanic activity that circles the Pacific Ocean.

Exploration 5
Teacher's Notes

Element of Surprise

Key Concepts	Elements in the same chemical group, or family, of the periodic table of elements have similar chemical properties. Predictions about the reactivity of an element may be based on the reactivity of another element in the same chemical group.
Summary	Mr. Fred Stamp of Pack & Mail, Inc. has to deliver some potentially dangerous chemicals to a remote research station on the continent of Antarctica. Because Mr. Stamp and his crew are going to be traveling via dog sled over and near water, he needs help determining the reactivity of each element to water. He has requested that each of the 12 elements that he will deliver be given a rating of extremely reactive, reactive, or not reactive to water. He will then use these ratings to construct the best possible transport containers for the elements.
Mission	Help to ensure that some potentially explosive chemical samples are safely delivered to the South Pole.
Solution	By testing the samples in the laboratory and by using their knowledge that elements in the same chemical group have similar chemical properties, students can determine the reactivity of the 12 samples. Krypton, like neon, radon, and xenon, is not reactive; magnesium, like calcium, barium, and strontium, is reactive; and sodium, like potassium, cesium, and rubidium, is extremely reactive.
Background	Shipping elements to a research team in Antarctica is not such a far-fetched idea. Antarctica is the continent about which we know the least. This vast region of unknowns lures scientists from all over the world to make new discoveries and to study everything from astrophysics to microbial ecology. In fact, the Antarctic Environmental Protection Act of 1996 designated Antarctica as a "natural reserve devoted to peace and science."
	Located almost completely within the Antarctic circle, Antarctica contains 90 percent of the world's ice. Because the ice sheets reflect most of the sun's heat back into the Earth's atmosphere, the South Pole region has an annual mean temperature of –49°C. Despite the extreme cold, scientists gather regularly at a number of antarctic research stations. At the South Pole Station, for example, astronomers from around the world study galaxy and star formation while astrophysicists keep an eye on the ozone layer. The McMurdo Dry Valleys are another hot spot for research on Antarctica. This unusual expanse of ice-free land, with its mountain ranges, meltwater streams, and arid terrain, draws botanists, geochemists, biologists, ecologists, and other scientists.

Exploration 5 Teacher's Notes, continued

Teaching Strategies

One likely outcome of this Exploration is that students will have a greater understanding of the fact that elements in the same chemical group of the periodic table share similar chemical properties. Students are more likely to stay focused on this project if they are reminded that in a real-life situation, an incorrect prediction about the reactivity of the elements could lead to serious injury or even death. As always, the students will find a good deal of helpful information in the CD-ROM articles, so be sure that students have read them carefully before reporting their findings to Mr. Stamp. If students have difficulty with this Exploration, encourage them to focus on what the tested chemicals in each reactive group have in common. Then direct their attention to the periodic table, and ask them what is similar about the placement in the table of the nonreactive, reactive, and highly reactive groups of elements.

As an extension, suggest that students research the types of containers that are appropriate for storing reactive, nonreactive, and highly reactive chemicals. Encourage the students to present their results in the form of a visual display. You may also wish to extend the Exploration with further analysis of other families in the periodic table. Divide the class into small groups, and provide each group with a list of several chemicals from the same family of the periodic table. Also include a description of the chemical properties of one of the chemicals. Then challenge the groups to make predictions about the chemical properties of the other chemicals in the list, based on the one description.

Bibliography for Teachers

Tocci, Salvatore, and Claudia Viehland. *Chemistry: Visualizing Matter.* Austin, TX: Holt, Rinehart and Winston, 1996.

Bibliography for Students

Cobb, Vicki. *Chemically Active: Experiments You Can Do at Home.* Philadelphia, PA: Lippincott, 1990.

Loesching, Louis V. *Simple Chemistry Experiments with Everyday Materials.* New York City, NY: Sterling Publishing Co., Inc., 1991.

Other Media

Chemistry: The Periodic Table and Periodicity
Film, videotape, and videodisc
Coronet/MTI Film & Video
P.O. Box 2649
Columbus, OH 43216
800-777-8100

Interested students may also find relevant information about elements on the Internet. Suggest that students use keywords such as the following to conduct their search: *chemistry, chemicals, chemical changes, elements,* and *reactivity.* Students may also wish to access interesting facts about *Antarctica* by exploring the Internet.

Name _____ Date _____ Class _____

Exploration 5
Worksheet

Element of Surprise

1. Mr. Stamp needs your help. Describe your assignment.

   _____
   _____
   _____
   _____
   _____

2. What materials are available in Dr. Labcoat's lab to help you complete your assignment?

   _____
   _____
   _____
   _____

3. Describe what you will do to test each element's reactivity to water.

   _____
   _____
   _____
   _____

4. Record your findings about each sample in the spaces that follow.

   a. barium: _____
   _____
   _____
   _____

   b. calcium: _____
   _____
   _____
   _____

40 SCIENCEPLUS INTERACTIVE EXPLORATIONS TEACHER'S GUIDE • LEVEL GREEN

Name _____ Date _____ Class _____

Exploration 5 Worksheet, continued

   c. cesium: _____

_____

_____

_____

   d. neon: _____

_____

_____

_____

   e. potassium: _____

_____

_____

_____

   f. radon: _____

_____

_____

_____

   g. rubidium: _____

_____

_____

_____

   h. strontium: _____

_____

_____

_____

   i. xenon: _____

_____

_____

_____

EXPLORATION 5 • ELEMENT OF SURPRISE

Name _____ Date _____ Class _____

Exploration 5 Worksheet, continued

5. What additional information do you need to complete your assignment (to determine the reactivity of krypton, magnesium, and sodium to water)?

_____
_____
_____
_____

6. Now that you know the reactivity of each of the 12 elements, how do you think Mr. Stamp should pack the chemicals when preparing to deliver them to Antarctica?

_____
_____
_____
_____
_____
_____
_____
_____
_____
_____
_____

Record your conclusions in the fax to Mr. Stamp.

Name _____ Date _____ Class _____

Exploration 5
Fax Form

FAX

To: Mr. Fred Stamp (FAX 011-619-555-7669)

From:

Date:

Subject: Chemical Properties of Elements

Select the appropriate classification for each of the following chemicals:

CHEMICAL REACTIVITY WITH WATER

CHEMICAL	EXTREMELY REACTIVE	REACTIVE	NOT REACTIVE
BARIUM			
CALCIUM			
CESIUM			
NEON			
POTASSIUM			
RADON			
RUBIDIUM			
STRONTIUM			
XENON			

Please utilize the above information to predict the chemical reactivity of the following chemicals:

CHEMICAL	EXTREMELY REACTIVE	REACTIVE	NOT REACTIVE
KRYPTON			
MAGNESIUM			
SODIUM			

How did the periodic table help you to answer Mr. Stamp's questions?

_____
_____
_____
_____

EXPLORATION 5 • ELEMENT OF SURPRISE   43

Exploration 5
CD-ROM Articles

Element of Surprise

The following articles can also be found by accessing the computer graphic of the CD-ROM for Exploration 5:

- *The Periodic Table of the Elements*
- *Properties of Matter*
- *Antarctica*

The Periodic Table of the Elements

How It Began

In the 1800s, scientists were struggling with a problem. Many elements had been discovered and analyzed, but scientists could not agree on the best way to organize the elements. Dmitri Mendeleev, a Russian chemist, proposed a solution that would later become our modern periodic table of the elements.

In 1869, Mendeleev arranged the elements according to their atomic masses. The atomic mass of an element equals the sum of the masses of the protons and neutrons in one atom of the element. He organized the elements in rows, with each row containing elements with similar properties. When Mendeleev created the periodic table, he predicted the future discovery of certain elements and their properties based on gaps that remained in the chart. Over time, scientists were able to fill in the chart. As they did, they found that most of Mendeleev's predictions were correct.

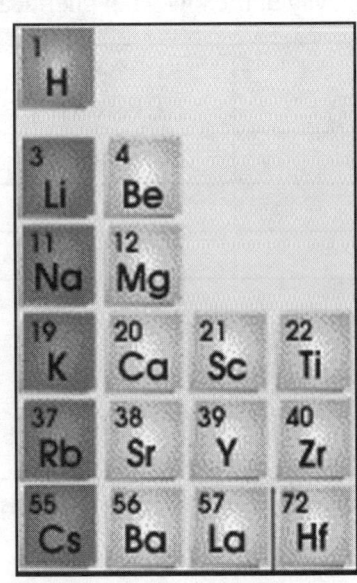

In the modern periodic table, atoms are arranged by atomic number—the number of protons in the nucleus of one atom of the element. Instead of organizing elements with similar properties in horizontal rows, the modern table organizes elements in columns. Scientists continue to modify the chart. And although all of the naturally occurring elements have been discovered, analyzed, and named, scientists can artificially create atoms with higher atomic numbers. These can then be analyzed, named, and placed on the chart.

Organizing the Elements—Trends in the Table

The modern periodic table summarizes a great deal of information about the elements. Each square on the table lists the name and symbol for the element, the atomic number, and the average atomic mass. In addition, the placement of each element on the table indicates some of the element's properties.

Elements are arranged by their increasing atomic numbers in the periodic table of the elements.

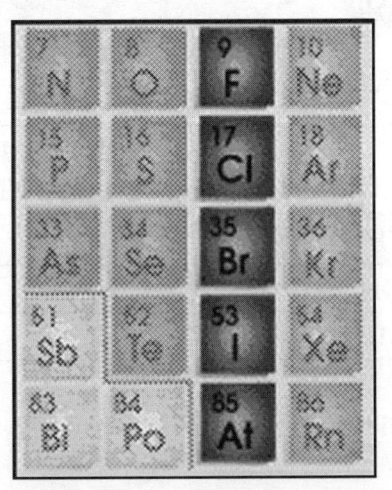

Elements in the same families have similar properties.

The atomic number establishes the order of the elements, but it is the number of electrons that determines the common properties of **groups** of elements. Groups are arranged in vertical columns on the table.

44 SCIENCEPLUS INTERACTIVE EXPLORATIONS TEACHER'S GUIDE • LEVEL GREEN

An element's group is determined by the number of electrons in the unfilled shell, or energy level, of the atom. The first shell can contain up to 2 electrons. The second and subsequent shells can contain up to 8 electrons. For example, hydrogen has an atomic number of 1. It has 1 electron in its unfilled shell. Lithium has an atomic number of 3. It has 2 electrons in its first shell and 1 electron in its unfilled shell. Because both hydrogen and lithium have a single electron in the unfilled shell, they belong to the same group. Elements in the same group share similar reactive properties. Elements that have their electron shells filled belong to a group called **noble gases.** These gases are sometimes classified as inert, or not reactive.

A Host of Elements

The periodic table lists over 100 different known elements. The descriptions here will give you an idea of some of the many differences between elements.

Barium
Atomic number: 56
Physical properties: Soft, silvery white metal
Barium compounds are often used when a doctor administers an X ray. The patient swallows a barium compound before having the X ray. The compound absorbs the X rays and reveals a clear picture of the digestive tract.

Calcium
Atomic number: 20
Physical properties: Moderately hard, silvery metal

Calcium is found in bones and shells of living organisms.

Calcium makes up about 3 percent of the Earth's crust. It is a primary component in bones, shells, and many rocks, including limestone and gypsum.

Cesium
Atomic number: 55
Physical properties: Soft, silvery white metal; liquid at room temperature
Cesium reacts quickly when exposed to light, so it is used in photocells. Scientists also use cesium to measure radiation from cosmic rays and nuclear particles.

Krypton
Atomic number: 36
Physical properties: Whitish and gaseous
Krypton is often used with other gases, such as neon, in fluorescent lamps and neon signs. It produces a bright, orange-red glow when an electric current is passed through it.

Magnesium
Atomic number: 12
Physical properties: Moderately hard, silvery metal
Because magnesium burns with a dazzling white light, it is used in fireworks, flares, and flashbulbs. Magnesium is also a part of the chemical chlorophyll, which allows plants to make food. It is an important component in certain medicines, fertilizers, and cements.

Neon
Atomic number: 10
Physical properties: Colorless and gaseous
Neon is an extremely rare element found in the Earth's atmosphere. It glows with a bright orange-red light when an electric current passes through it.

Potassium
Atomic number: 19
Physical properties: Soft, silvery white metal
Potassium is usually combined with other elements. It is a major component of the Earth's crust and plays an important role in the nervous system of many animals. Extremely explosive, potassium is used in matches and fireworks. It also occurs in combination with other elements in soaps, fertilizers, and dyes.

Exploration 5 CD-ROM articles, continued

Radon

Atomic number: 86
Physical properties: Colorless, radioactive, and gaseous

Radon is formed from the disintegration of the element radium. Because it produces radiation, it is very dangerous to life-forms. Radon has been discovered in homes built over soil and rock that have high concentrations of radium.

Rubidium

Atomic number: 37
Physical properties: Soft, silvery white

Rubidium ignites in air and reacts violently in water. Like cesium, its high reactivity makes it useful in photocells.

Sodium

Atomic number: 11
Physical properties: Soft, silver-white metal; malleable

The sixth most abundant element in the Earth's crust, sodium is found in both soil and water. It regulates water and nerve function in living cells. Sodium and chlorine form table salt, a common chemical. Sodium is also an important ingredient in baking powder, lye, soap, fertilizers, and many other products.

Strontium

Atomic number: 38
Physical properties: Soft, silvery white metal

Strontium ignites in air and produces a brilliant red flame. It is often used in fireworks and flares. Strontium-90 is a form of strontium that is radioactive and dangerous to living things.

Xenon

Atomic number: 54
Physical properties: Colorless, odorless, and gaseous

Xenon is found in tiny quantities in the atmosphere. It is used in certain lamps and light bulbs because it can produce a bright light.

Properties of Matter

Physical and Chemical Properties of Matter

A physical property is a property of matter that can be observed without changing the composition of the matter. Mass, hardness, boiling point, melting point, color, and texture are all physical properties. Physical properties are often easy to observe. Consider the physical properties of pure water in a glass. In its liquid state, water is clear, odorless, and has a measurable mass. You do not have to change the composition of the water to make these observations. If you wanted to determine the water's freezing or boiling point, you still would not change the water's composition.

This is not the case with a chemical property. A chemical property is any property of matter that describes how one kind of matter interacts with other kinds of matter. When a piece of iron is left in the rain, for example, it reacts with the water to form iron oxide, or rust. When iron rusts, a chemical change occurs. The very composition of the iron changes. Because this chemical change occurs, a chemical property of iron is that it reacts with water to form iron oxide, or rust.

Each element has a unique set of physical and chemical properties. However, elements in the same group tend to share certain chemical properties. This occurs because many chemical properties are determined by the reactivity of the element. The reactivity of an element is largely based upon the number of electrons in the element's unfilled electron shell.

Exploration 5 CD-ROM articles, continued

Antarctica

Antarctica is the Earth's most southern continent and largest icecap.

A History of the Continent

After studying fossils and geological formations, some scientists have speculated that Antarctica once had a very different climate and position on Earth. Over 200 million years ago, Antarctica may have been a part of Gondwanaland, a large, warm continent that joined all seven existing continents.

The continent of Antarctica, an area that measures approximately 13 million square kilometers, is almost entirely covered by a thick sheet of ice. The ice sheet was created by the buildup of snow over millions of years, and it contains over 70 percent of the world's freshwater supply. As the ice piles up, it turns into glaciers and ice rivers that flow from the continent to the sea. Often, large glaciers break off and float out to sea until they melt in warmer waters. A vast current called the Antarctic Circumpolar Current moves the cold waters into the rest of the world's oceans.

Exploration

Antarctica was the last continent to be discovered and explored. Many people wanted to reach the South Pole, which lies on the interior of the frozen land. The first expeditions that landed on the continent's coast were recorded in 1895. By 1901, British explorer Robert F. Scott began the first expedition to reach the geographic South Pole. He failed in his attempts and was beaten to the location by Norwegian Roald Amundsen in 1911. These early explorers moved across the continent using dog sleds, sails, and other primitive resources. Many of these adventurers died in the frozen climate.

By the late 1920s, airplanes made conquering the difficult terrain easier. In November 1929, American Richard Byrd flew over the geographic South Pole. He continued exploring Antarctica by air and on land. By the late 1950s, several nations had joined together to establish major scientific research centers at the geographic South Pole and around the continent.

The South Poles

Antarctica actually contains three points known as south poles. The geographic South Pole lies at a latitude of 90 degrees south of the Earth's equator. The magnetic south pole is the location that compasses point toward. It is located to the north and east of the geographic South Pole. The geomagnetic south pole is the location of the Southern Hemisphere's auroras, which are brightly colored lights that appear in the sky. It lies between the geographic and magnetic poles.

Exploration 6
Teacher's Notes

The Generation Gap

Key Concepts Alternative sources of electricity, such as wind and solar power, can be less expensive than coal, oil, and other common sources of electricity. A number of factors can affect the viability of wind-generated electricity.

Summary Ms. Wendy Powers of EcoCabins, Inc. would like to know if it would be cost-effective to use the Electroprop wind turbine to provide electricity for a small log home in the San Francisco area. The turbines cost $14,560 and should not need repairs for 20 years. Ms. Powers would like to know how long it will take for the turbine to pay for itself.

Mission Test a wind turbine to determine if it will lower the overall energy costs for a small log home.

Solution If installed in the San Francisco area, the Electroprop wind turbine would save the homeowner an average of $1456 a year and would pay for itself in 10 years. Since the turbine should not need repairs for 20 years, the turbine should be considered highly cost-effective.

Background Fossil fuel supplies are limited, and most people agree that we are using these resources much faster than they can be replaced by nature. For this reason, fossil fuels are considered to be nonrenewable resources. In addition to being limited, fossil fuels have other drawbacks. Oil spills, strip mining for coal, and exhaust from power plants and motor vehicles can all have negative effects on our environment. As a result, many nations today are focusing on developing new energy sources, making the most of the ones we already have, and using every form of energy as efficiently as possible. Many energy-wise plans focus on renewable resources, resources that are continually produced. Some renewable energy resources, such as wind and sunlight, are so abundant that they are considered to be inexhaustible.

Tremendous energy is contained in the wind. This energy can be captured with a turbine, which is connected to an electric generator. The technology of wind turbines is well developed, and the cost of wind-generated electricity is lower than that produced by some other sources. The primary disadvantage of wind energy is that few regions have winds strong or consistent enough to make wind turbines economical. In addition, large, commercial wind turbines may not be attractive additions to the landscape. At full speed they can be very noisy. The blades of large wind turbines can also interfere with microwave communications. While wind energy may never be a major source of energy, it is a practical energy alternative in some areas.

Exploration 6 Teacher's Notes, continued

Teaching Strategies

In order to complete this Exploration successfully, students must have a clear understanding of the necessary mathematical calculations. These calculations are outlined in the CD-ROM articles. You may wish to review the CD-ROM articles with students who are having difficulty determining the correct procedure. Make sure students understand that they must compare the total energy savings over a long period of time with the cost of installing the wind turbine. If the wind turbine pays for itself before it needs repairs, then the turbine can be considered cost-effective.

As an extension of this Exploration, you may wish to invite an energy-resource expert from a local college, university, or scientific institution to speak to the class about current energy-efficient technologies.

Bibliography for Teachers

Brower, Michael. *Cool Energy: Renewable Solutions to Environmental Problems.* Revised edition. Cambridge, MA: MIT Press, 1992.

Burmeister, George, and Frank Kreith. *Energy Management and Conservation.* Washington, DC: National Conference of State Legislatures, 1993.

Bibliography for Students

Cozic, Charles, ed. *Current Controversies: Energy Alternatives.* San Diego, CA: Greenhaven Press, 1994.

Inhaber, Herbert, and Harry Saunders. "Road to Nowhere," *The Sciences*, Nov.–Dec. 1994, pp. 20–25.

Other Media

Electric Bill
Software (Apple II family)
Queue, Inc.
338 Commerce Dr.
Fairfield, CT 06432
203-335-0906
800-232-2224

Power Struggle
Videotape
Bullfrog Films
P.O. Box 149
Oley, PA 19547
215-779-8226
800-543-3764

Interested students may also find relevant information about alternative energy sources on the Internet. Suggest that students use keywords such as the following to conduct their search: *energy, alternative sources of energy, fossil fuels, conservation, electricity,* and *wind turbines.* Students may also be interested in searching for information about *energy-efficient homes.*

Name _____ Date _____ Class _____

Exploration 6
Worksheet

The Generation Gap

1. Wendy Powers is a home builder who is considering a plan to make her homes more efficient. What has she asked you to do to help her?

   _____
   _____
   _____
   _____

2. Dr. Labcoat has set up a system that enables you to test the energy output of the wind turbine at eight different speed settings. Run the tests, and record your results below.

Meters per second	Kilowatt-hours	Time-lapse indicator

3. What is the value of the above information?

   _____
   _____
   _____
   _____
   _____

Name _____ Date _____ Class _____

Exploration 6 Worksheet, continued

4. What other information will you need to complete your task?

   _____
   _____
   _____
   _____
   _____
   _____

5. Use the lab resources to find this information. You can record your notes here.

   _____
   _____
   _____
   _____
   _____
   _____
   _____
   _____
   _____
   _____
   _____

6. How will you calculate the amount of money a wind turbine can save a homeowner over the course of a year?

   _____
   _____
   _____
   _____

EXPLORATION 6 • THE GENERATION GAP 51

Name _____ Date _____ Class _____

Exploration 6 Worksheet, continued

7. Use the table below to record the energy output of the Electroprop wind turbine and your calculations of the savings it will bring the homeowner.

Wind speed in meters per second (m/s)	Energy output over 7 days in kilowatt-hours (kWh)	Savings per year ($)	Years until Electroprop wind turbine has paid for itself

Record your conclusions in the fax to Ms. Powers.

Name _____ Date _____ Class _____

Exploration 6
Fax Form

FAX

To: Ms. Wendy Powers (FAX 415-555-2766)

From:

Date:

Subject: Wind-Energy Economics

Is it cost-effective to use the Electroprop to generate energy in the San Francisco area? Why or why not?

_____

_____

_____

_____

✂ ✂

For Internal Use Only

Please answer the following questions for my laboratory records. Scientists must always keep good records. *Dr. Crystal Labcoat*

Approximately how much money would the Electroprop save a San Francisco homeowner in an average year?

$4	$28	$140	$400	$1460	$2800

Approximately how many years would it take for the Electroprop to pay for itself?

1	5	10	16	28	250

How many years would it take for the Electroprop to pay for itself if the average wind speed in the San Francisco area were each of the following:

8 m/s? _____

5 m/s? _____

2 m/s? _____

EXPLORATION 6 • THE GENERATION GAP 53

The Generation Gap

The following articles can also be found by accessing the computer graphic of the CD-ROM for Exploration 6:

- *The Cost of Electricity*
- *Wind Turbines*

Exploration 6
CD-ROM Articles

The Cost of Electricity

Calculating Electricity Costs

Electricity usage is measured by a unit called the *kilowatt-hour*, abbreviated *kWh*. The average American household uses approximately 1000 kWh of electricity each month. Utility companies charge each household for the electricity used, and the cost of electricity varies from city to city. Consider the cost of electricity in five American cities.

Average Monthly Electricity Usage and Cost

City	Cost per kWh
Atlanta, GA	$0.06
Chicago, IL	$0.05
Dallas, TX	$0.05
Houston, TX	$0.09
San Francisco, CA	$0.14

What is the average cost of electricity per month for a household in each city? To find out, multiply the number of kilowatt-hours of electricity used by the cost per kilowatt-hour. For example, the cost of electricity in Atlanta is six cents per kilowatt-hour. Since the average household uses 1000 kWh of electricity in one month, the average cost of electricity per month is:

1000 kWh × $0.06 = $60.00

```
ELECTRIC BILL
IN 29 DAYS YOU USED          249 KWH
  READ DATE            METER #0000
  03/29                      5121
  02/29                      4872
  DIFFERENCE                 249 KWH
  RATE CALCULATION:          249 KWH
  RESIDENTIAL SERVICE
  CUSTOMER CHARGE:           $6.00
    ENERGY:   $.03550/KWH    $8.84
    FUEL:     $.01583/KWH     3.84
  SUBTOTAL ELECTRIC CHARGES  $18.78
    SALES TAX                  .19

  TOTAL COST FOR
  ELECTRIC SERVICE:          $18.97
```

A residential customer pays for each kilowatt-hour of electrical energy.

How to Read an Electric Bill

The charge for your electricity may be included in a bill along with charges for water, garbage collection, and other services. Or charges for electricity may come in a separate bill. Either way, the bill will show how much electricity was used in a given amount of time and the cost of the electricity.

Consider the following example of an electric bill.

```
IN 29 DAYS YOU USED           831 kWh
READ DATE              METER # 00328450
05/07/96                        1463
04/08/96                         632
DIFFERENCE                       831

RATE CALCULATION
CUSTOMER CHARGE                $6.00
831 kWh AT $0.09/kWh           74.79

SUBTOTAL ELECTRIC CHARGES     $80.79

SALES TAX                       0.81

TOTAL COST FOR
ELECTRIC SERVICE              $81.60

FOR THIS 29-DAY PERIOD, YOUR
AVERAGE DAILY COST FOR
ELECTRIC SERVICE WAS           $2.81
```

The first part of the bill shows you that on April 8, the meter read 632 kWh. On May 7, the meter read 1463 kWh. The difference, 831 kWh, is the total number of kWh used from April 8 to May 7.

The next part of this bill shows the actual charges. First, it shows a service charge of $6.00. This charge covers the cost of reading the meter and processing the bill. The cost of the electricity follows. The number of kilowatt-hours is multiplied by the city's charge per kilowatt-hour—nine cents in this example. The service charge and the electricity charge are subtotaled, and a sales tax is computed.

The last part of this bill shows the total cost for electric service and the average cost per day for electricity.

In this sample electric bill, the charge per kilowatt-hour represents the cost of the actual fuel used to produce the electricity and the cost of operations and maintenance at the utility company. In some cities, these charges may be represented separately. In addition, some utility companies may charge an increased rate if a household uses more than a certain number of kilowatt-hours, or the charge per kilowatt hour may vary depending on what time of day the electricity is used.

How to Lower Your Household's Electric Bill

Electric bills can cost your household a large amount of money each month. However, you can do several things to help reduce the amount of electricity used and thus lower your household's total energy bill.

Use less hot water. In many homes, water is heated using electricity. Keep showers short, and use a cold-water rinse cycle when using the washing machine.

Take charge of the thermostat. Electricity is often used to heat and cool homes. In the winter, lower the thermostat and wear a sweater indoors. In the summer, raise the thermostat and use fans to keep cool.

Turn off the lights. Leaving the lights on when they are not in use wastes electricity. Lights also generate heat and can add to cooling costs in the summer.

Turn off appliances. Turn off the TV, radio, and other appliances when no one is using them.

Wind Turbines

What Is Wind?

The movement of air across the Earth's surface is known as *wind*. Winds are set in motion by the uneven heating of the Earth's surface, and they tend to blow from regions of higher to lower air pressure. Winds are divided into two types—*local* and *planetary*.

Consider an oceanside setting. During the day, the land heats up faster than the water, and the warm land heats the air above it. As the hot air rises, the pressure it puts on the land decreases. The cool air over the water is heavy and dense, thus creating an area of high pressure. This cool, heavy air over the water then moves to the low-pressure area over the land. At night, the reverse occurs. Land cools faster than water. Thus, the air moves from the high-pressure area over the land to the low-pressure area over the ocean. Winds created by this process are known as *local winds*. Local winds are set in motion by the surface features of a particular area, such as bodies of water, ice formations, and land formations.

Pressure differences across the globe also create winds called *planetary winds*. Air rises in warm regions along the equator, creating areas of low air pressure. After a time this air stops rising and drifts toward the polar regions. Most of this air cools and sinks before it reaches the poles, creating areas of high pressure. The pattern of rising and sinking air creates pressure belts along the Earth's surface. Planetary winds flow between pressure belts, from areas of high pressure to areas of low pressure.

What Is a Wind Turbine?

A wind turbine is a device that harnesses the energy of wind and converts it into electrical energy. In a wind turbine, wind turns large propellers that are attached to a generator. As the generator spins, it produces electricity that can be carried along power lines.

A wind turbine uses renewable wind energy to make electricity.

A typical wind turbine has a tower, blades, a generator and cables, and an electronic control system. The tower raises the blades to a point where the wind is strong and blows steadily. A typical turbine has two or three blades that span about 18 m. The generator produces electricity from the spinning blades, and the electricity travels down the tower along the cables. At the base of the turbine, a control system monitors the electricity production.

A wind power plant, or wind farm, has large clusters of wind turbines. These are often owned by private businesses that sell the electricity produced to utility companies. Because wind turbines are large, a successful wind farm requires a large amount of land along with appropriate wind conditions.

Advantages and Disadvantages of Using Wind as an Energy Source

There are several advantages to using wind as an energy source. First and foremost, wind is a renewable resource because it is produced by the sun. The sun heats the Earth's surface unevenly and causes disturbances in the atmosphere, forming winds. As long as the sun shines, the Earth will have winds. Another advantage of wind turbines is that they can use the wind without adding pollutants to the air, water, or soil.

Also, wind-generated electricity is often less expensive than that produced by other sources. Modern wind farms can produce electricity at just five cents per kilowatt-hour. Traditional power plants produce electricity at five to six cents per kilowatt-hour, require more expensive maintenance, and create more pollution.

Harnessing wind energy also has a number of disadvantages. To be effective, wind turbines require steady winds that do not fall below 3.1 m/s. The most efficient production of energy occurs when wind speed averages about 4.5 m/s. Many regions lack winds that are consistent and strong enough to make a wind turbine system affordable. In addition, most wind farms can only produce electricity about 25 percent of the time; traditional power plants, such as coal-operated plants, can operate around the clock. Also, commercial wind generators can be noisy, and their large propellers can make the landscape unattractive.

An anemometer measures wind speed.

Wind Speeds in the United States

The chart below shows the average wind speed in seven United States cities. As you can see, average wind speed varies around the nation.

City	Wind speed (m/s)
Boston, MA	5.0
Las Vegas, NV	4.0
New Orleans, LA	3.0
Oklahoma City, OK	6.0
Richmond, VA	3.0
San Francisco, CA	6.0
St. Louis, MO	4.0

Most wind turbines operate in the Midwest, the Northeast, and California. In fact, wind energy produces about 1 percent of California's electricity, making California the leading state for harnessing wind resources.

Since wind speed varies, homes that use wind power for their electricity must also have a second energy source. On days when the wind speed is insufficient, the second energy source provides electricity. When the wind turbine produces more electricity than the household needs, the excess can be sold to a utility company. Sometimes the sale of this excess energy can save a household 50 to 90 percent on its monthly electric bill.

Wind Turbines—Are They Worth It?

Wind turbines can be a safe and affordable method of generating electricity. To determine whether a wind turbine will be an energy source for a household, the following factors must be considered:

- average wind speed in the region
- number of kilowatt-hours produced by the wind turbine per week
- cost of kilowatt-hours charged by the local utility company
- cost of wind turbine

Here is one example of a household that found that money could be saved by using a wind turbine. In Nevada, where the Hioes live, the average wind speed is 4.0 m/s, which produces about 65 kWh of electricity in one week. The local utility company charges 13 cents per kWh. On average, the Hioes use 1000 kWh per month, or 12,000 kWh per year.

Approximate yearly savings using a wind turbine:

Electricity from wind turbine:
  65 kWh × 52 weeks = 3380 kWh per year

Average cost of electricity from the local utility company:
  $0.13 per kWh

Savings per year:
  3380 kWh × $0.13 = $439.40

Using one wind turbine to generate electricity, the Hioes could save approximately $439.40 in one year.

How long would it take for the wind turbine to pay for itself?

If the initial cost of the Hioes' wind turbine is $7000, it would take approximately 16 years for the wind turbine to pay for itself.

Exploration 7
Teacher's Notes

Teach It While It's Hot!

Key Concepts	Heat and temperature are different concepts. Graphs can be used as a visual aid to make these and other scientific concepts more readily understandable.
Summary	In the past, Mr. Kelvin McCool, a science teacher at J. P. Joule Middle School, has experienced some problems teaching his students the difference between temperature and heat. He has asked Dr. Labcoat and her assistants to help him by preparing a demonstration and some graphs to illustrate the characteristics of both temperature and heat.
Mission	Help Mr. McCool teach the relationship between heat and temperature to his class of middle-school students.
Solution	Heat refers to the total energy of motion of the atoms and molecules in a substance, while temperature is a measure of the average energy of motion of the atoms and molecules in a substance. A volume of water twice the size of another volume of water must contain twice as much heat to maintain the same temperature as the smaller volume.
Background	Temperature is an average measure of the movement of the constantly vibrating particles that make up all matter. The hotter a substance gets, the more energy its particles have, and the faster the particles move. The colder a substance gets, the less energy its particles have, and the slower the particles move. Scientists theorize that all movement of particles stops at absolute zero (–273°C). By using laser devices in laboratories, scientists have cooled matter to within a millionth of a degree of absolute zero.
	Working with extremely cold temperatures has resulted in the development of some important technologies. Cryosurgery allows doctors to use low temperatures to seal off blood vessels during an operation. Superconductors are materials that have been cooled to super-low temperatures so that they can conduct electricity without resistance. The opportunities made possible by a perfectly efficient conductor are nearly endless.
Teaching Strategies	Successfully completing this Exploration depends on thoroughly researching the CD-ROM articles and correctly interpreting the graph of the results. Students must recognize that the amount of heat needed to raise the temperature of a mass of water a specific amount is proportional to that mass. Because the density of water = 1 g/1 mL, this proportion can also be discussed in terms of volume.

Exploration 7 Teacher's Notes, continued

For instance, a 300 mL volume of water will require three times as much heat as a 100 mL volume to raise its temperature the same amount. You can use the graph in the Exploration to emphasize the differences in the amount of heat required to raise different volumes of water to specific temperatures. Encourage students to study the graph of their results carefully before attempting to complete the fax form. You may wish to review the CD-ROM articles on graphing with students who are having difficulty reading the graph.

As an extension of this Exploration, have students use what they discovered to create a graph that shows the amount of heat required to raise the temperature of 400 mL, 500 mL, and 600 mL of water from 20°C to 100°C. You may also wish to suggest that students discuss the advantages of understanding the relationship between heat and temperature. For instance, ask students how a cook might change his or her approach to heating food based on a knowledge of this relationship. A larger (more massive) pan or pot would require more heat energy to heat up than one that is less massive. For example, it would take longer to boil a certain volume of water in a 2 kg pan than in a 1 kg pan.

Bibliography for Teachers

Cuevas, Mapi M., and William G. Lamb. *Physical Science*. Austin, TX: Holt, Rinehart and Winston, 1994.

Bibliography for Students

Maury, Jean-Pierre. *Heat & Cold*. Hauppauge, NY: Barron's Educational Series, Inc., 1989.

Wood, Robert W. *Physics for Kids—Forty-Nine Easy Experiments with Heat*. Blue Ridge Summit, PA: TAB Books Inc., 1989.

Other Media

Heat and Energy Transfer
Film and videotape
Coronet/MTI
P.O. Box 2649
Columbus, OH 43216
800-777-8100

Heat: Molecules in Motion
Videodisc
AIMS Media
9710 DeSoto Ave.
Chatsworth, CA 91311-4409
818-773-4300

Heat, Temperature, and the Properties of Matter
Film and Videotape
Coronet/MTI
P.O. Box 2649
Columbus, OH 43216
800-777-8100

In addition to the above resources, interested students may find relevant information about heat and temperature on the Internet. Suggest that students conduct a search with keywords such as the following: *heat and temperature, heat and energy, Newtonian universe,* and *James Joule*.

EXPLORATION 7 • TEACH IT WHILE IT'S HOT!

Name _____ Date _____ Class _____

Exploration 7
Worksheet

Teach It While It's Hot!

1. What has Dr. Labcoat asked you to do to help Mr. McCool?
   _____
   _____
   _____

2. What information would be helpful to know before you begin your investigation?
   _____
   _____
   _____

3. Where do you think you could find this information?
   _____
   _____

4. What happens when heat energy is applied to a beaker of water?
   _____
   _____
   _____

5. Record your observations as each beaker (quantity) of water is placed on the ring stand.
   a. green (100 mL)
   _____
   _____
   _____

   b. red (200 mL)
   _____
   _____
   _____

60 SCIENCEPLUS INTERACTIVE EXPLORATIONS TEACHER'S GUIDE • LEVEL GREEN

Exploration 7 Worksheet, continued

   c. blue (300 mL)

6. How can you calculate the amount of heat required to increase the temperature of 600 mL of water from 20°C to 100°C?

7. Why did Dr. Labcoat provide you with three different quantities of water?

Name _____ Date _____ Class _____

Exploration 7 Worksheet, continued

8. Use the graph as well as your knowledge of temperature and heat to describe what this demonstration shows.

_____
_____
_____
_____
_____
_____
_____
_____
_____
_____
_____
_____
_____
_____
_____

9. Based on what you've learned during this activity, would you recommend this demonstration to Mr. McCool? Why or why not?

_____
_____
_____
_____
_____
_____
_____

Record your conclusions in the fax to Mr. McCool.

SCIENCEPLUS INTERACTIVE EXPLORATIONS TEACHER'S GUIDE • LEVEL GREEN

Name _____ Date _____ Class _____

Exploration 7
Fax Form

FAX

To: Mr. Kelvin McCool (FAX 512-555-4328)

From:

Date:

Subject: Teaching Recommendations

What relationship is represented by your graph?

_____

_____

_____

_____

_____

_____

Please use your data to determine the answers to the following questions:

Which beaker contains the most heat energy at 100°C?	GREEN	RED	BLUE
Approximately how much heat would have to be added to increase the temperature of 600 mL of water from 20°C to 100°C?	100,000 joules	200,000 joules	300,000 joules
What is the approximate temperature of each sample of water when the amount of heat energy added is 30,000 joules? GREEN: ☐ RED: ☐ BLUE: ☐			

EXPLORATION 7 • TEACH IT WHILE IT'S HOT!   63

Name _____ Date _____ Class _____

Exploration 7 Fax Form, continued

Please write and answer one essay question that will help Mr. McCool's students understand the relationship between temperature and heat.

Teach It While It's Hot!

The following articles can also be found by accessing the computer graphic of the CD-ROM for Exploration 7:

- *Temperature and Heat*
- *Graphs*

Temperature and Heat

What's the Difference?

Compare the following two definitions:

Temperature—the measure of the average kinetic energy of the motion of atoms and molecules in a substance

Heat—the internal energy of a substance that is due to the random movement of the atoms and molecules in a substance

If you compare these definitions closely, you will see that temperature refers to the *average kinetic energy* of a substance while heat refers to the *total kinetic energy* of a substance. To explore this difference further, imagine that you have two mugs of hot cocoa, both at the same temperature and both made of the same material. Would you say that they both contain the same amount of heat energy? The answer is yes if they both have the same mass.

But what if one mug were larger and had more mass than the other mug? Would the more massive mug have the same amount of heat as the smaller one, even if they were both at the same temperature? The answer is no. Even though the molecules in each mug have approximately the same kinetic energy, the larger mug has more molecules, and therefore has more heat energy.

Mass Is Related to Heat

Now think about making some eggs and potatoes to go with the cocoa. First you must heat two frying pans made of the same material. Let's say that one pan has a mass of 1 kg and the other pan has a mass of 2 kg. If you placed the two pans on identical burners set to the same temperature, the heavier frying pan would take twice as long to get hot. That's because the more massive the object, the more heat energy it takes to raise its temperature.

Suppose that you wanted to raise the temperature of each frying pan by 1°C. To do this, the 2 kg frying pan would require exactly twice as much heat as the 1 kg frying pan would require. So the effect of mass on temperature can be summed up this way: when the mass of a substance is doubled, it takes twice as much heat to increase the temperature of that mass by 1°C.

The larger frying pan takes longer to get hot.

The larger mug has more cocoa and more heat.

Measuring Heat

In the metric system, heat energy is measured in *joules*. The joule is named after James Prescott Joule, a nineteenth-century scientist who theorized that mechanical energy and heat energy were different forms of the same thing. A joule (J) is approximately equal to the amount of mechanical energy required to raise a 100-gram mass 1 meter. If measuring heat energy, it takes 4.19 J of heat to raise the temperature of 1 g of

Exploration 7 CD-ROM articles, continued

water by 1°C. Heat can also be measured in calories—one calorie equals the amount of heat needed to increase the temperature of 1 g of water by 1°C. So 1 calorie = 4.19 J.

Measuring Temperature

Thermometers are used to measure temperature. Many thermometers work on the principle that matter expands when heated and contracts when cooled. A glass thermometer, for example, has a narrow tube that is filled with either mercury or alcohol. These liquids are used because they do not freeze or boil easily. When heated, the mercury or alcohol expands inside the tube. And since the rate of expansion for these liquids is predictable, the glass tube can be calibrated with degree marks that correspond to a temperature scale.

The temperature scales most often used are the Celsius and Fahrenheit scales. In the metric system, water freezes at 0°Celsius (°C) and boils at 100°C. On the Fahrenheit scale, water freezes at 32°Fahrenheit (°F) and boils at 212°F. The Fahrenheit scale is named after the German physicist Gabriel Fahrenheit, who in 1714 invented the glass mercury thermometer.

Specific Heat

Raising a substance's temperature always means increasing the heat energy of the substance; it always takes energy to raise temperature. But some substances need less energy to heat up. The term *specific heat* describes the amount of heat needed to raise the temperature of 1 g of a substance by 1°C. If you have 1 g of water, you have to add 1 calorie of heat to increase the temperature by 1°C. But most substances need a lot less energy to get hot. For instance, you need only 0.22 calories of heat to raise the temperature of 1 g of aluminum by 1°C.

Few substances have a specific heat as great as that of water. Water can absorb a large amount of heat energy without a large change in its temperature. In fact, the ocean (which feels cool when we go swimming in it) actually holds an enormous amount of heat energy. Its large mass combined with the high specific heat of water keep the ocean temperature relatively cool.

Graphs

Types of Graphs

Often scientists use various types of graphs to illustrate information. Check out the following three examples of graphs. They all show information that would not be as clear if described by words only.

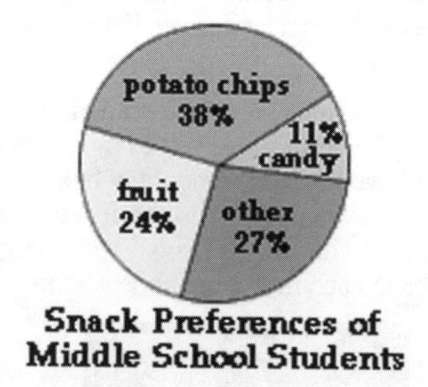

Snack Preferences of Middle School Students

potato chips 38%
11% candy
fruit 24%
other 27%

Exploration 7 CD-ROM articles, continued

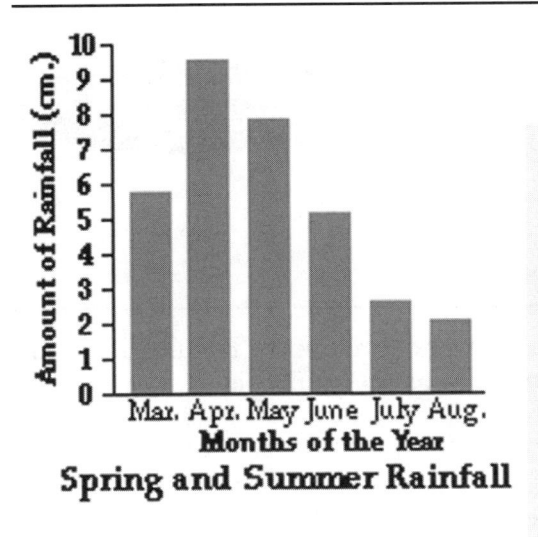

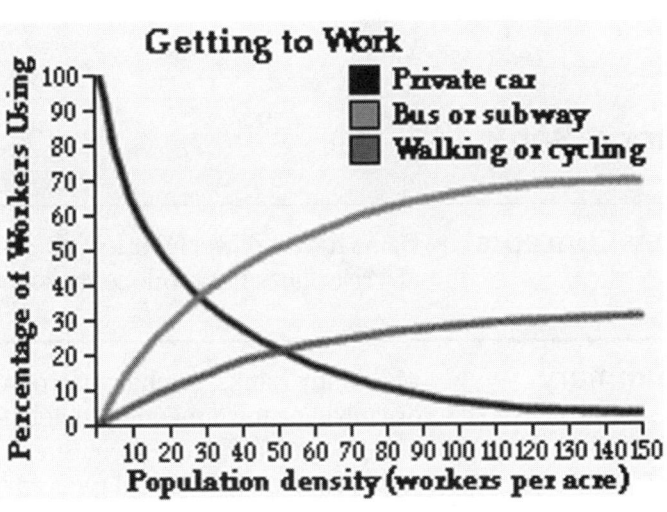

Parts of a Graph

The diagram below shows the parts of a graph. Note that both the *x*- and *y*-axes increase in regular intervals. That makes reading the graph easier. A good graph always has clear labels and a title.

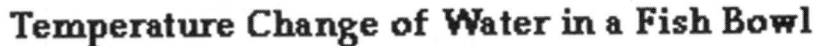

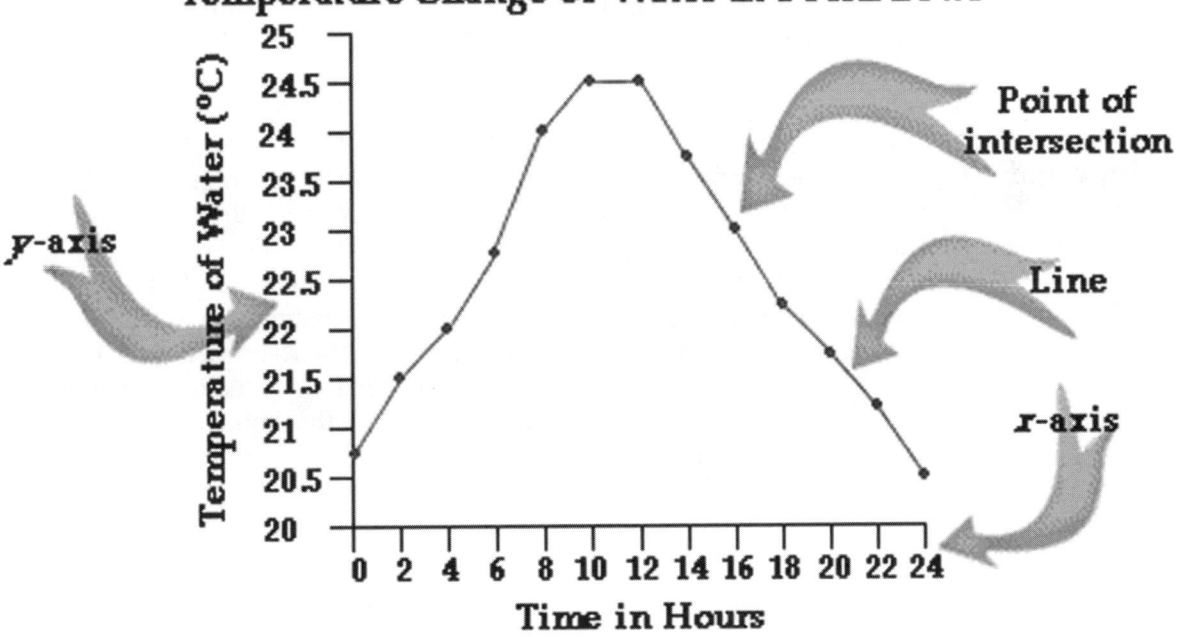

EXPLORATION 7 • TEACH IT WHILE IT'S HOT! 67

Flood Bank

Key Concepts	Dams have a powerful effect on natural river environments. Regulated flooding of a river does not produce the same results as natural flooding.
Summary	Ms. Sandy Banks, as chairman of her local environmental-impact committee, is organizing a town meeting to debate whether a dam should be built in the community to create a reservoir. She needs to know what long-term effects a dam would have on the natural river environment.
Mission	Predict the effects of a dam on the natural flow of a river, its geological formations, and its ecosystem.
Solution	Dams block the natural flow of a river and limit the amount of sediment that the river carries downstream. This lack of sediment can drastically affect ecosystems that depend on the sediment in the river as the basis of their food chains. Additionally, regulated flooding does not produce the same scouring of the river bottom that natural flooding does. The resulting changes in the landscape threaten the organisms that inhabit the river.
Background	Altering the flow of river water can change the ecosystem of not only the local river environment but also the environment many kilometers downstream. Large, fan-shaped deposits at the mouth of a river are called deltas, and they are home to a wide variety of fish, crustaceans, and mollusks. Because the ocean is continually sweeping portions of the sediment out to sea, a delta depends on new large deposits of sediment from the river each year. The Mississippi River, for example, deposits 2 million tons of sediment into the Gulf of Mexico each year. A dam upstream would limit the amount of sediment that the river would be able to carry downstream, and the delta and its ecosystem could potentially disappear.

Exploration 8 Teacher's Notes, continued

Teaching Strategies

Successfully completing this Exploration requires that students research the CD-ROM articles thoroughly. Encourage students to use the Notepad function to take notes from the articles.

As an extension of this Exploration, you may wish to have students research natural alterations to the courses of rivers and streams, such as dams built by beavers. Ask students to find out which plants and animals depend on these natural water regulators and how removing beaver dams can disrupt ecosystems by increasing the rate of the river's flow.

Bibliography for Teachers

"Science and Technology: The Beautiful and the Dammed." *The Economist*, 322 (7752): March 28, 1992, p. 93.

Sides, Hampton. "Let There Be High Water." *Outside*, 21 (7): July 1996, pp. 38–41, 104–106.

Bibliography for Students

Kaufman, Jeffrey S., Robert C. Knott, and Lincoln Bergman. *River Cutters*. Berkeley, CA: Lawrence Hall of Science, 1990.

Wuerthner, George. "Dammed River, Doomed Mollusks." *Defenders,* 67 (3): May/June 1992, pp. 12–13.

Other Media

Erosion and Weathering: Looking at the Land
Erosion: Leveling the Land
   Two videotapes
   Britannica
   310 S. Michigan Ave.
   Chicago, IL 60604-9839
   800-554-9862

Interested students may find relevant information about dams and their effects on the environment by accessing the Internet. Suggest that students conduct a search using keywords such as the following: *river dams, flooding,* and *altering ecosystems.*

EXPLORATION 8 • FLOOD BANK   69

Name _____ Date _____ Class _____

Exploration 8
Worksheet

Flood Bank

1. Ms. Sandy Banks is the chairperson of her local environmental-impact committee. What has she asked you to do to help her?

   _____
   _____
   _____
   _____

2. How can the simulation that Dr. Labcoat has provided help you with your investigation?

   _____
   _____
   _____
   _____
   _____
   _____

3. Conduct your research, recording your notes in the space provided below.

   _____
   _____
   _____
   _____
   _____
   _____
   _____
   _____
   _____
   _____

SCIENCEPLUS INTERACTIVE EXPLORATIONS TEACHER'S GUIDE • LEVEL GREEN

Name _____ Date _____ Class _____

Exploration 8 Worksheet, continued

4. Use the table below to record the effects of both types of water flow at varying flow rates.

Type of flow	Low	Medium	High
Natural flow			
Regulated flow			

5. Compare the results of your stream-table observations with the information you discovered in your research. Which environmental and geological effects of dams are not reproducible by a stream table?

_____
_____
_____
_____
_____
_____

Record your conclusions in the fax to Ms. Banks.

EXPLORATION 8 • FLOOD BANK 71

Name _____ Date _____ Class _____

Exploration 8
Fax Form

FAX

To: Ms. Sandy Banks (FAX 520-555-7239)

From:

Date:

Subject: Possible effects of a dam on the natural river environment

What effect does a dam have on the natural river environment?

_____
_____
_____
_____
_____
_____
_____
_____
_____
_____
_____
_____
_____
_____
_____
_____
_____

72 SCIENCEPLUS INTERACTIVE EXPLORATIONS TEACHER'S GUIDE • LEVEL GREEN

Name _____ Date _____ Class _____

Exploration 8 Fax Form, continued

Which volume of water flow has the greatest impact on the formation of geologic features?

| Low | Medium | High |

Is it possible to maintain a healthy river environment downstream from a dam?

| YES | NO |

Why or why not?

_____
_____
_____
_____
_____
_____
_____
_____
_____
_____
_____
_____
_____
_____
_____
_____
_____
_____
_____
_____

EXPLORATION 8 • FLOOD BANK

Flood Bank

The following articles can also be found by accessing the computer graphic of the CD-ROM for Exploration 8:

- Anatomy of a River
- Questions About Controlling a River
- A Case Study

Exploration 8
CD-ROM Articles

Anatomy of a River

From Top to Bottom

Rivers flow from areas of higher elevation to areas of lower elevation. Several rivers get their start in the high elevations of the Rocky Mountains. Rainwater and snowmelt run down the slopes of the mountains, forming small channels that eventually become streams and rivers. These streams and rivers empty into larger rivers such as the Colorado River, which runs west from the Rocky Mountains. The rivers and streams that empty into larger rivers are called *tributaries*. For example, the Illinois River is a tributary of the mighty Mississippi River. Most river systems will continue to flow downhill until their water reaches the sea. The point where a river empties into the sea is called the river's *mouth*.

As the water of a river flows, some of the land along the river and along the river bottom is washed downstream. As the flowing water erodes the riverbanks and river bottom, it becomes rich in *sediment* consisting of small rocks, soil, and other particles. The sediment carried by the river is often called the river's *load*. The load of a river increases with the river's size and flow rate; a large, swift river causes more erosion and carries more sediment than a small, slow-moving river.

A river's load is not always constant. For example, a river will deposit some of its load wherever the river water slows down—such as on the inside of curves, in areas where the river widens, or where the riverbanks are less steep. This process is called *deposition*.

Growth and Aging

Rivers change over time. Compare the differences between a young, mature, and old river.

Young river: A young river is characterized by steep banks, narrow or V-shaped valleys, and rapid erosion. Young rivers often form steep gorges or canyons.

Mature river: A mature river is characterized by gently sloping banks, wide valleys, and moderate erosion.

Old river: An old river is characterized by low banks, wide and flat valleys, and little erosion. Wide curves called *meanders* mark the shape of an old river. And although an old river does not cause much erosion, it does carry and distribute large amounts of sediment that it receives from its tributaries.

Some rivers can be young at their source, mature in the middle, and old near their mouth. As the river ages, it spreads over the *flood plain*. The flood plain is the flat, low-lying area that is covered by flowing water during times of high water volume.

Seasonal Flow and Flooding

The flow rate of a river can change from season to season and even from day to day. In the spring, for example, mountain streams swell with water as the snow melts. This causes a surge in the river's flow rate, even far downstream.

Floods occur when heavy rain or melting snow enters a river system causing the river to overflow its banks. A flood causes a tremendous increase in erosion. A flood can move boulders as well as tons of sediment. A flood can even change the course of a river.

When the flood water slows down, it deposits sediment along the river bank and the flood plain. This sediment is very rich in nutrients. And when it is deposited in the flood plain, it serves to maintain healthy sandbars, which in turn provide spawning grounds for many species of fish and other organisms.

Even though floods appear to be very damaging to the environment, they also serve to maintain the natural ecosystems in and around the river. On the other hand, human-made structures on a flood plain are very much at risk during floods. Because floods are so destructive to anything in their path, people try to control the flow of the water by deepening the riverbeds, by building walls along the riverbanks, and by constructing dams.

Questions About Controlling a River

What Does a Dam Do?
Dams enable us to control the flow of water in a river. A dam built across a river creates an artificial lake, called a *reservoir*, behind the dam. Water can be released from the reservoir at a rate that prevents flooding downstream. Water held in the reservoir is often used for drinking, for recreational purposes, and for irrigating farms.

In addition to the uses listed above, some dams are equipped with electrical generators that operate as the water flows over large turbines. These dams generate electricity and are often referred to as hydroelectric dams.

How Does a Dam Affect the River's Environment?
Because a dam blocks the natural flow of a river, it limits the erosion and deposition of sediment by the river. So instead of collecting and carrying nutrient-rich sediment downstream, the river drops the sediment at the bottom of the reservoir. As a result, the ecosystems downstream are deprived of the sediment that is often the basis of their food chains.

The water released through a dam is also cooler than naturally flowing river water. Many organisms cannot adapt to this cooler water, and die off. Again, this may break a natural food chain and cause serious damage to the river ecosystem.

Dams also prevent flood waters from periodically scouring the river bottom. As a result, debris settles on the river's bottom and creates barriers in front of inlets where fish would normally spawn. Where flood waters once carved out the river's banks and created new sandbanks farther downstream, vegetation takes root. These changes in the landscape threaten the organisms that live in and along the river system.

Can We Control the Flow to Protect the River System?
Dams are very disruptive to the natural river environment. Even if water could flow through the dam at the same rate it would normally flow, much of the river's sediment would remain on the bottom of the reservoir. Controlled flooding could help keep the river clear of debris and could help distribute some of the sediment below the dam, but sediment above the dam would still need to be moved somehow from behind the dam to areas downstream. Also, some method would need to be devised to increase the temperature of the water before it affected the ecosystems downstream.

A Case Study

The Colorado River and the Glen Canyon Dam
Before construction of the Glen Canyon Dam in 1963, spring floods of over 93,000 cubic feet per second rushed down Glen Canyon, eventually flooding through the Grand Canyon. However, since 1963, these spring floods of the Colorado River have been controlled by Glen Canyon Dam.

The spring floods once cleaned the riverbed, built beaches along the banks, and removed vegetation that clogged the river's path. Since 1963, however, vegetation has thickened, nutrient levels have declined, sediment has been lost, water temperatures have dropped, and natural sandbars have disappeared. Many of the species that lived in the Colorado River before the dam was built, such as the humpbacked chub, are now endangered. Others are extinct. New species, such as the rainbow trout, have been introduced to the river and have disrupted the ecosystem's food web.

After more than 10 years of data collection and planning, scientists and government officials began an experiment. On March 27, 1996, a controlled flood was released from Glen Canyon Dam. Each day for one week, 45,000 cubic feet per second of water cascaded from the dam. It was hoped that this controlled flooding would restore some of the ecological conditions that existed in the Colorado River before the dam was built. Early results indicate that although the natural sediment is still being held back by the dam, some of the beaches and sandbars and some of the natural habitats have been temporarily restored.

Answer Keys

SciencePlus Interactive Explorations Teacher's Guide

Contents

Exploration 1 78
Exploration 2 81
Exploration 3 86
Exploration 4 90
Exploration 5 93
Exploration 6 97
Exploration 7 101
Exploration 8 106

Answer Key for Exploration 1

Exploration 1 Worksheet

Something's Fishy

1. What kinds of problems is Mr. McMullet having with his African Cichlids?
   Answers should include details about the condition of the fish, such as their color, their low-energy behavior, and the fact that some are dying.

   (Recommended 5 pts.)

2. What are five variables that might be affecting the African Cichlids?
   a. **Frequency of feeding**
   b. **Presence of ornamental driftwood**
   c. **Variation in temperature**
   d. **Age of the filter**
   e. **Amount of light** *(Recommended 10 pts.)*

3. What will you use for a control as you conduct your investigations?
   One of the two tanks, left undisturbed with settings that remain constant

   (Recommended 5 pts.)

4. Why is this control necessary?
   To establish a comparison with the experimental tank, in which variables can be isolated and the results of changing those variables can be observed *(Recommended 10 pts.)*

5. Would it be better to test one variable at a time or several variables at once? Why?
   It is better to test one variable at a time, because this is the only way of isolating the variable responsible for the change in the cichlids. *(Recommended 10 pts.)*

6. Form a hypothesis for each of the experiments you conduct.
   Hypothesis 1: **Hypotheses should include a reasonable estimation of how each variable might or might not affect the health of cichlids.** *(Recommended 30 pts.)*

78 SCIENCEPLUS INTERACTIVE EXPLORATIONS TEACHER'S GUIDE • LEVEL GREEN

Answer Key for Exploration 1, continued

Hypothesis 2: _____

Hypothesis 3: _____

Hypothesis 4: _____

Hypothesis 5: _____

7. Record your observations as you investigate each hypothesis.

Hypothesis	Observations
1	**Observations should include any changes in the health, behavior, and color of the cichlids.** *(Recommended 20 pts.)*
2	
3	
4	
5	

8. Were your experiments faulty in any way? If so, what steps did you take to correct your experiments?

Errors may include failure to conduct research before beginning an experiment, trying to change the control tank, and testing more than one variable at a time.

(Recommended 10 pts.)

Record your conclusions in the fax to Mr. McMullet.

Answer Key for Exploration 1, continued

Exploration 1
Fax Form

FAX

To: Mr. Ray McMullet (FAX 512-555-8633)

From:

Date:

Subject: African Cichlids

What is your recommendation? **Remove the ornamental driftwood from your tanks; it is harming the cichlids.** *(Recommended 10 pts.)*

For Internal Use Only

Please answer the following questions for my laboratory records. Scientists must always keep good records. — *Dr. Crystal Labcoat*

During your experiments, which ONE of the following changes had a positive effect on the fish?

EXPERIMENTAL VARIABLES

☐ FEED FISH ☐ INCREASE TEMPERATURE
☐ TURN LIGHT OFF ☐ CHANGE FILTER
☒ REMOVE ORNAMENTAL DRIFTWOOD

(Recommended scoring: 50 points for above answer; 30 points for that response plus another; and 0 points for not indicating the above response.)

Please explain why the African Cichlids responded to the above change.

African Cichlids require water that is alkaline. The driftwood that was added to the tanks was slowly leaching tannins into the water. The tannins caused the water to be too acidic for the cichlids. Removing the driftwood corrected the pH level in the tank.

(Recommended 25 pts.)

What effect did the change that you made have on the fish?

The fish became active and colorful. *(Recommended 15 pts.)*

80 SCIENCEPLUS INTERACTIVE EXPLORATIONS TEACHER'S GUIDE • LEVEL GREEN

Answer Key for Exploration 2

Exploration 2
Worksheet

Shut Your Trap!

1. What are Ms. Lily N. Lotus and the Bogs Are Beautiful Appreciation Society concerned about?

   Answers should include a summary of the society's concerns, including a description of why the flytraps are disappearing from the wild. Examples may include poaching and a shrinking natural habitat. *(Recommended 5 pts.)*

2. What are three variables that may be affecting the growth of the Venus' flytraps?

   a. Light

   b. Humidity

   c. Plant food   *(Recommended 10 pts.)*

3. What will you use for a control in your investigations?

   One of the two terrariums, left undisturbed with settings that remain constant

   (Recommended 5 pts.)

4. There are 24 possible variable settings for the experimental terrarium. However, you have only enough flies to conduct 10 experiments. What steps can you take to make sure that you find a solution before you run out of flies?

   The order in which variables are tested can quickly reduce the number of experiments necessary. For instance, testing the effect of the plant food first will eliminate half of the possible number of experiments. *(Recommended 20 pts.)*

5. Form a hypothesis for how each variable affects the growth of the Venus' flytraps.

   Hypothesis 1: Hypotheses should include a reasonable estimation of how variables may or may not combine to affect the growth of the flytraps. *(Recommended 20 pts.)*

   Hypothesis 2: _____

   Hypothesis 3: _____

EXPLORATION 2 • ANSWER KEY   81

Answer Key for Exploration 2, continued

6. Record your observations in the table below as you investigate each hypothesis.

Plant food	Humidity	Hours of light	Observations
Yes	0%	5	**Observations should include specific changes in the growth, health, or blooming of the fly-traps.** *(Recommended 30 pts.)*
		10	
		15	
		20	
	25%	5	
		10	
		15	
		20	
	50%	5	
		10	
		15	
		20	
No	0%	5	
		10	
		15	
		20	
	25%	5	
		10	
		15	
		20	
	50%	5	
		10	
		15	
		20	

Answer Key for Exploration 2, continued

7. Were your experiments faulty in any way? If so, what steps did you take to correct them?

   Errors may include failure to conduct research before beginning the experiments, incorrectly setting the variables on the control terrarium, or testing variables at random.

   (Recommended 10 pts.)

Record your conclusions in the fax to Lily N. Lotus.

Answer Key for Exploration 2, continued

Exploration 2
Fax Form

FAX

To: Lily N. Lotus (FAX 910-555-5657)

From:

Date:

Subject: Optimal Growing Conditions for Venus' Flytraps

What is your recommendation? **Optimal growing conditions for the Venus' flytrap include 15 hours of light, 50% humidity, and no plant food.** *(Recommended 5 pts.)*

For Internal Use Only

Please answer the following questions for my laboratory records. Scientists must always keep good records. Dr. Crystal Labcoat

During your experiments, which values proved to be optimal for the Venus' flytrap?

Optimal Values (select one per row)   **EXPERIMENTAL VARIABLES**

5	10	**15**	20	HOURS OF LIGHT PER DAY
0	25	**50**		PERCENT HUMIDITY
YES	**NO**			ADD PLANT FOOD?

(Recommended scoring: 50 points if optimal values are indicated as shown; 30 points for 10 hours of light, 50% humidity, and no plant food; and 30 points for 20 hours of light, 50% humidity, and no plant food. All other answers should receive 0 points.)

What effect did the hours of light per day have on the plants? Why?

The Venus' flytrap produces a healthy seedpod in 90 days when exposed to 15 hours of light per day.

(Recommended 10 pts.)

What effect did the percent humidity have on the plants? Why?

The Venus' flytrap produces a healthy seedpod in 90 days when exposed to 50% humidity.

(Recommended 10 pts.)

84   SCIENCEPLUS INTERACTIVE EXPLORATIONS TEACHER'S GUIDE • LEVEL GREEN

What effect did the plant food have on the plants? Why?

Plant food cannot be given to the Venus' flytrap because the plant has adapted to the nutrient-poor soil of the Carolina bogs. Since the plant receives nitrogen from insects, the extra nutrients provided by plant food will kill the plant. *(Recommended 25 pts.)*

Answer Key for Exploration 3

Exploration 3
Worksheet

Scope It Out!

1. What does Dr. Viola Russ need to know about the ancient microorganisms?

   Dr. Russ needs to know the classification of the ancient microorganisms and information about the likely role they played in the life of the ancient bee. *(Recommended 5 pts.)*

2. What does Dr. Russ intend to do with the results?

   She will use this information to produce new and more effective antibiotics.

   (Recommended 5 pts.)

3. What will you use to conduct your investigation?

   Making a visual comparison of the ancient microorganisms and the modern microorganisms is the only way to make an identification, so the slides should be carefully analyzed under the microscope. Answers should also emphasize researching the CD-ROM articles for clues about the ancient microorganisms' function. *(Recommended 10 pts.)*

4. What do the ancient microorganisms look like under the microscope?

   The ancient microorganisms are reddish pink, single celled, and have an oblong shape.

   (Recommended 10 pts.)

86 SCIENCEPLUS INTERACTIVE EXPLORATIONS TEACHER'S GUIDE • LEVEL GREEN

Answer Key for Exploration 3, continued

5. Use the table below to record your observations of each slide of microorganisms. Make sure that you write out the name of each microorganism in the left-hand column.

Protista	Observations
1. Euglena	Transparent green; single celled; one flagellum
2. Paramecium	Transparent blue; single celled; with cilia
3. Amoeba	Reddish brown; single celled; uses cytoplasm as pseudopodia
4. Algae, dinoflagellate	Gray and pink; two flagella

Monera	Observations
5. Round-shaped bacteria, *Staphylococcus aureus*	Red and green; single celled; round shape
6. Rod-shaped bacteria, *Escherichia coli*	Orange with green spots; single celled; oblong shape
7. Spiral-shaped bacteria, *Spirillum*	Transparent purple; single celled; spiral shape

Fungi	Observations
8. Mildew, *Peronospora manschuriea*	Black, curving fibers
9. Mold, *Rhizopus stolonifer*	Black spots on yellowish, hairlike fibers
10. Yeast, *Saccaromyces cerevisiae*	Blue, round, interconnected bubbles

(Recommended scoring: 2 points for each correct name and 2 points for each appropriate description.)

EXPLORATION 3 • ANSWER KEY   87

Answer Key for Exploration 3, continued

6. Which modern microorganisms look the most like the ancient microorganisms?

   The rod-shaped bacteria of the kingdom Monera share the same basic shape as the ancient microorganisms. *(Recommended 10 pts.)*

7. How might you classify the ancient microorganisms and find out more about their likely role in the life of the ancient bee?

   Information can be obtained from the CD-ROM articles, as well as from outside references such as books, periodicals, and other materials about microorganisms. Consulting an expert from a local college or university could also be helpful. *(Recommended 20 pts.)*

Record your conclusions in the fax to Dr. Russ.

Answer Key for Exploration 3, continued

Exploration 3
Fax Form

FAX

To: Dr. Viola Russ (FAX 805-555-2266)

From:

Date:

Subject: Ancient microorganism classification and probable function

What are the ancient microorganism's classification and function? *The ancient micro-organism is a type of rod-shaped bacteria that belongs to the kingdom Monera. The organism probably aided the bee in digestion and in fighting disease.*

(Recommended 25 pts.)

✂·

For Internal Use Only

Please answer the following questions for my laboratory records. Scientists must always keep good records. Dr. Crystal Labcoat

Which of the following may be used to classify the ancient microorganisms? Place an X in the left-hand column beside the correct answer(s).

KINGDOM		PROTISTA	X	MONERA		FUNGI
		Euglena		Round-shaped bacteria		Mildew
		Paramecium	X	Rod-shaped bacteria		Mold
		Amoeba		Spiral-shaped bacteria		Yeast
		Algae				

(Recommended scoring: 50 points if both kingdom Monera and rod-shaped bacteria are selected; 30 points if kingdom Monera and round-shaped or spiral-shaped bacteria are selected; 30 points if only kingdom Monera or only rod-shaped bacteria is selected; and 0 points for all other answers.)

What role did this microorganism most likely play in the life of the ancient bee?

Like Bacillus sphaericus, the ancient microorganism probably aided the bee in digestion and in fighting disease.   *(Recommended 25 pts.)*

Answer Key for Exploration 4

Exploration 4 Worksheet

What's the Matter?

1. What problem does Dr. Stokes need you to help him solve?

   The tip of an instrument for measuring lava temperature has melted, and Dr. Stokes needs some help determining what metal to use as a replacement. *(Recommended 10 pts.)*

2. Dr. Labcoat has gathered a set of metal samples for you to analyze. List the name of each metal and record its melting point and boiling point in the following data chart:

Metal	Name of metal	Melting point (°C)	Boiling point (°C)
Cu	**Copper**	**1083**	**2567**
Sn	**Tin**	**232**	**2270**
Pt	**Platinum**	**1722**	**3827**
Ti	**Titanium**	**1660**	**3287**
W	**Tungsten**	**3410**	**5660**
Al	**Aluminum**	**660**	**2467**

(Recommended 30 pts.)

3. How might the information in the table be useful to you in solving Dr. Stokes's problem?

   Dr. Stokes's instrument will have to be repaired with a material that is very heat resistant since it will be inserted into hot lava. It is important, therefore, to determine the temperature limits of likely materials. *(Recommended 10 pts.)*

4. What additional information could help you solve the problem?

   Answers should include information such as the following: the temperature of molten lava, the availability and expense of each metal, and how difficult it is to obtain each metal.

   (Recommended 10 pts.)

Answer Key for Exploration 4, continued

5. How might you find this information?

   The CD-ROM articles are an excellent source of such information. In addition, outside resources such as books, scientific articles, or an expert at a local college, university, or scientific institution may be consulted. *(Recommended 10 pts.)*

6. Record helpful data here, continuing on the back of the page if necessary.

   Answers should reflect a careful review of the CD-ROM articles.

   (Recommended 10 pts.)

7. When you have finished, evaluate the procedure that you used to complete this Exploration. What would change about your procedure? Did you perform any activities that were not useful to you? If so, which ones?

   Answers will vary but should include a clear reflection of the steps taken to complete this Exploration. Answers may include the fact that determining the mass and volume of each metal was not useful in solving Dr. Stokes's problem. *(Recommended 10 pts.)*

8. How could you improve your procedure?

   Answers will vary but could include the fact that carefully planning an experiment before proceeding is an effective way to avoid problems. *(Recommended 10 pts.)*

Record your conclusions in the fax to Dr. Stokes.

Answer Key for Exploration 4, continued

Exploration 4
Fax Form

FAX

To: John Stokes, Ph.D. (FAX 080-555-9822)

From:

Date:

Subject: Metal recommendation

What is your recommendation? **Use titanium to replace the tip of the lava analyzer.**

(Recommended 20 pts.)

For Internal Use Only

Please answer the following questions for my laboratory records. Scientists must always keep good records. *Dr. Crystal Labcoat*

Please indicate your metal selection here: **Titanium**

(Recommended scoring: 50 points for above answer; 30 points for tungsten or platinum; and 0 points for copper, aluminum, or tin.)

How do the particles of this metal behave during the following phases:

solid? **The particles are packed closely together in a rigid form until about 1659°C.**

(Recommended 10 pts.)

liquid? **At about 1660°C, the particles begin to separate, vibrating against each other and flowing in a random pattern.** *(Recommended 10 pts.)*

gas? **At about 3287°C, the particles begin to bounce off each other and into the air in a random way.** *(Recommended 10 pts.)*

Answer Key for Exploration 5

Exploration 5
Worksheet

Element of Surprise

1. Mr. Stamp needs your help. Describe your assignment.

   The assignment is to determine the chemical reactivity of 12 elements to water so that Mr. Stamp can construct the best possible transport containers for the elements and deliver them safely to Antarctica. *(Recommended 10 pts.)*

2. What materials are available in Dr. Labcoat's lab to help you complete your assignment?

   There are samples of 9 of the 12 elements, a beaker of water, a pair of tweezers, a pipette, an eyedropper, and a chemical storage rack. *(Recommended 10 pts.)*

3. Describe what you will do to test each element's reactivity to water.

   A sample of each element should be added to the beaker of water, and then observations can be made about any reaction that takes place. *(Recommended 10 pts.)*

4. Record your findings about each sample in the spaces that follow. *(Recommended 40 pts.)*

   a. barium: When tested, the water bubbles quickly and turns cloudy.
   Answers to a-i should include a description of each sample's reaction to water. Students may also have included additional information about each element from the CD-ROM articles.

   b. calcium: When tested, the water bubbles quickly and turns cloudy.

EXPLORATION 5 • ANSWER KEY 93

Answer Key for Exploration 5, continued

c. cesium: When tested, the water turns purple and there is a slightly explosive reaction above the surface of the water.

d. neon: When tested, the water bubbles briefly but remains clear.

e. potassium: When tested, the sample bubbles on the surface of the water before creating a smoking reaction above the surface. The water changes color.

f. radon: When tested, the water bubbles briefly but remains clear.

g. rubidium: When tested, the top portion of the water turns purple, and there is a slightly explosive reaction on the surface.

h. strontium: When tested, the sample sinks to the bottom of the beaker and bubbles rise quickly toward the surface, changing the color of the water.

i. xenon: When tested, the water bubbles briefly but remains clear.

5. What additional information do you need to complete your assignment (to determine the reactivity of krypton, magnesium, and sodium to water)?

The periodic table of elements in Dr. Labcoat's lab and the information in the CD-ROM articles are needed to determine the reactivity of the remaining elements to water.

(Recommended 15 pts.)

6. Now that you know the reactivity of each of the 12 elements, how do you think Mr. Stamp should pack the chemicals when preparing to deliver them to Antarctica?

Sample answer: It is very important that neither the highly reactive nor the reactive elements come in contact with water. Therefore, the containers that hold the cesium, potassium, rubidium, sodium, barium, calcium, strontium, and magnesium samples should be watertight. The samples of neon, radon, xenon, and krypton will not be damaged if they come in contact with water, so it is less important for those containers to be watertight.

(Recommended 15 pts.)

Record your conclusions in the fax to Mr. Stamp.

Answer Key for Exploration 5, continued

Exploration 5
Fax Form

FAX

To: Mr. Fred Stamp (FAX 011-619-555-7669)

From:

Date:

Subject: Chemical Properties of Elements

Select the appropriate classification for each of the following chemicals:

CHEMICAL REACTIVITY WITH WATER

CHEMICAL	EXTREMELY REACTIVE	REACTIVE	NOT REACTIVE
BARIUM		X	
CALCIUM		X	
CESIUM	X		
NEON			X
POTASSIUM	X		
RADON			X
RUBIDIUM	X		
STRONTIUM		X	
XENON			X

(Recommended scoring: Each correct answer is worth 2 points, for a total of 18 points possible.)

Please utilize the above information to predict the chemical reactivity of the following chemicals:

CHEMICAL	EXTREMELY REACTIVE	REACTIVE	NOT REACTIVE
KRYPTON			X
MAGNESIUM		X	
SODIUM	X		

(Recommended scoring: 50 points for three correct answers; 30 points for 1 or 2 correct answers; 0 points for no correct answers.)

How did the periodic table help you to answer Mr. Stamp's questions?

Elements in the same column of the periodic table belong to the same chemical group and have similar chemical properties. Knowing how one element in a group reacts with water can be used to predict the reactivity of other elements in that same group.

(Recommended 32 pts.)

96 SCIENCEPLUS INTERACTIVE EXPLORATIONS TEACHER'S GUIDE • LEVEL GREEN

Answer Key for Exploration 6

Exploration 6 Worksheet

The Generation Gap

1. Wendy Powers is a home builder who is considering a plan to make her homes more efficient. What has she asked you to do to help her?

   Ms. Powers has asked for assistance in testing the Electroprop wind turbine to determine if it would be cost-effective to install in the San Francisco area. *(Recommended 4 pts.)*

2. Dr. Labcoat has set up a system that enables you to test the energy output of the wind turbine at eight different speed settings. Run the tests, and record your results below.

Meters per second	Kilowatt-hours	Time-lapse indicator
1	1	7 days
2	8	7 days
3	27	7 days
4	65	7 days
5	125	7 days
6	200	7 days
7	350	7 days
8	500	7 days

   (Recommended 16 pts.)

3. What is the value of the above information?

   It shows the amount of energy generated by several speeds of wind in a 7-day period of time. *(Recommended 10 pts.)*

Answer Key for Exploration 6, continued

4. What other information will you need to complete your task?

    The average wind speed in the San Francisco Bay area, the cost per kilowatt-hour of electricity in the San Francisco Bay area, how to calculate the amount of money saved by using a wind turbine, and how to calculate the amount of time before the wind turbine pays for itself

    (Recommended 10 pts.)

5. Use the lab resources to find this information. You can record your notes here.

    Answers should reflect a careful review of the CD-ROM articles.

    (Recommended 10 pts.)

6. How will you calculate the amount of money a wind turbine can save a homeowner over the course of a year?

    By multiplying the number of kilowatt-hours generated by the wind turbine by the cost per kilowatt-hour to find the amount of savings per week, and then by multiplying the weekly savings by 52 weeks to find the amount of savings per year

    (Recommended 14 pts.)

Answer Key for Exploration 6, continued

7. Use the table below to record the energy output of the Electroprop wind turbine and your calculations of the savings it will bring the homeowner.

Wind speed in meters per second (m/s)	Energy output over 7 days in kilowatt-hours (kWh)	Savings per year ($)	Years until Electroprop wind turbine has paid for itself
1	1	7.28	2000
2	8	58.24	250
3	27	196.56	75
4	65	473.20	31
5	125	910.00	16
6	200	1456.00	10
7	350	2548.00	6
8	500	3640.00	4

(Recommended 36 pts.)

Record your conclusions in the fax to Ms. Powers.

Answer Key for Exploration 6, continued

Exploration 6
Fax Form

FAX

To: Ms. Wendy Powers (FAX 415-555-2766)

From:

Date:

Subject: Wind-Energy Economics

Is it cost-effective to use the Electroprop to generate energy in the San Francisco area? Why or why not?

Yes, because the Electroprop wind turbine will pay for itself in 10 years. After this initial pay-back period, the turbine will save the homeowner approximately $1456 each year for 10 years, for a total savings of $14,560. *(Recommended 20 pts.)*

For Internal Use Only

Please answer the following questions for my laboratory records. Scientists must always keep good records. Dr. Crystal Labcoat

Approximately how much money would the Electroprop save a San Francisco homeowner in an average year?

| $4 | $28 | $140 | $400 | **$1460** | $2800 |

Approximately how many years would it take for the Electroprop to pay for itself?

| 1 | 5 | **10** | 16 | 28 | 250 |

(Recommended scoring: 50 points for two correct answers; 30 points for one correct answer; and 0 points for no correct answers.)

How many years would it take for the Electroprop to pay for itself if the average wind speed in the San Francisco area were each of the following:

8 m/s? **4 years**

5 m/s? **16 years**

2 m/s? **250 years**

(Recommended scoring: 10 points for each correct answer.)

100 SCIENCEPLUS INTERACTIVE EXPLORATIONS TEACHER'S GUIDE • LEVEL GREEN

Answer Key for Exploration 7

Exploration 7 Worksheet

Teach It While It's Hot!

1. What has Dr. Labcoat asked you to do to help Mr. McCool?

   To make sure that the demonstration she has prepared will be an effective teaching tool for Mr. McCool's lesson about the difference between temperature and heat

   (Recommended 5 pts.)

2. What information would be helpful to know before you begin your investigation?

   A familiarity with both temperature and heat as well as the units in which each is measured would be useful. *(Recommended 5 pts.)*

3. Where do you think you could find this information?

   In the CD-ROM articles or in a science textbook *(Recommended 5 pts.)*

4. What happens when heat energy is applied to a beaker of water?

   The temperature of the water increases. *(Recommended 5 pts.)*

5. Record your observations as each beaker (quantity) of water is placed on the ring stand.

   a. green (100 mL)

   The water begins to boil after 6 seconds. It takes about 34,000 J of heat energy to bring 100 mL of water from 20°C to 100°C. *(Recommended 10 pts.)*

   b. red (200 mL)

   The water begins to boil after 11 seconds. It takes about 67,000 J of heat energy to bring 200 mL of water from 20°C to 100°C. *(Recommended 10 pts.)*

Answer Key for Exploration 7, continued

c. blue (300 mL)

The water begins to boil after 17 seconds. It takes about 101,000 J of heat energy to bring 300 mL of water from 20°C to 100°C.   *(Recommended 10 pts.)*

6. How can you calculate the amount of heat required to increase the temperature of 600 mL of water from 20°C to 100°C?

Sample answer: There is a direct relationship between the mass of a substance and the amount of heat required to raise the temperature of that mass by a specific amount. Since a 600 mL beaker contains six times as much water as a 100 mL beaker, six times as many joules will be required to raise the temperature of 600 mL of water from 20°C to 100°C as are required to raise the temperature of 100 mL of water from 20°C to 100°C.

(Recommended 15 pts.)

7. Why did Dr. Labcoat provide you with three different quantities of water?

To show the direct relationship between the amount (mass) of water and the amount of heat energy needed to raise the water to a specific temperature. For example, it requires three times as much energy to raise 300 mL of water from 20°C to 100°C as it does to raise 100 mL of water from 20°C to 100°C.   *(Recommended 10 pts.)*

Answer Key for Exploration 7, continued

8. Use the graph as well as your knowledge of temperature and heat to describe what this demonstration shows.

   Sample answer: This demonstration shows that when heat energy is added to a substance, the temperature of that substance increases. That temperature is an average measure of how "hot" the substance is at that particular instant, whereas, the heat of the substance tells how much total energy the substance contains. The amount of heat required to raise the temperature of a substance to a specific temperature is directly related to the mass of the substance (in this case, the volume of the water).   *(Recommended 15 pts.)*

9. Based on what you've learned during this activity, would you recommend this demonstration to Mr. McCool? Why or why not?

   Answers will vary, but recommendations should be clear, concise, and well supported with examples. Sample response: Yes. This demonstration effectively shows that there is a direct relationship between the mass of a substance and the amount of heat required to raise the temperature of that mass by a specific amount. The visual demonstration and the graph present the information in a format that is understandable to middle-school students.

   (Recommended 10 pts.)

Record your conclusions in the fax to Mr. McCool.

EXPLORATION 7 • ANSWER KEY   103

Answer Key for Exploration 7, continued

Exploration 7
Fax Form

FAX

To: Mr. Kelvin McCool (FAX 512-555-4328)

From:

Date:

Subject: Teaching Recommendations

What relationship is represented by your graph?

The graph shows the relationship between change in temperature and amount of heat added. The three different lines on the graph represent the three different quantities of water and show the direct relationship between mass and the amount of heat required to cause a specified change in temperature. For example, if the mass of a sample of water is doubled, then the amount of heat required to raise the temperature of that sample of water from 20°C to 100°C will also double.

(Recommended 10 pts.)

Please use your data to determine the answers to the following questions:

Which beaker contains the most heat energy at 100°C?	GREEN	RED	**BLUE**
Approximately how much heat would have to be added to increase the temperature of 600 mL of water from 20°C to 100°C?	100,000 joules	**200,000 joules**	300,000 joules

(Recommended scoring: 50 points for two correct answers; 30 points for one correct answer; 0 points for no correct answers.)

What is the approximate temperature of each sample of water when the amount of heat energy added is 30,000 joules?

GREEN: **92°C**   RED: **56°C**   BLUE: **44°C**

(Recommended scoring: each correct answer is worth 5 points, for a total of 15 points.)

SCIENCEPLUS INTERACTIVE EXPLORATIONS TEACHER'S GUIDE • LEVEL GREEN

Please write and answer one essay question that will help Mr. McCool's students understand the relationship between temperature and heat.

Sample question: What is the difference between temperature and heat? Sample answer: Heat is the total amount of energy in a substance, and temperature is a measurement of the average amount of energy in a substance, or how "hot" a substance is. A 100 mL beaker of water and a 500 mL beaker of water that are the same temperature have different amounts of heat energy. The more massive amount of water will contain more heat energy.

(Recommended 25 pts.)

Answer Key for Exploration 8

Exploration 8 Worksheet

Flood Bank

1. Ms. Sandy Banks is the chairperson of her local environmental-impact committee. What has she asked you to do to help her?

   Ms. Banks needs to know whether building a dam to create a reservoir will have long-term geologic and environmental effects on the river downstream.

   (Recommended 15 pts.)

2. How can the simulation that Dr. Labcoat has provided help you with your investigation?

   The simulation provides a means of visually comparing the geological effects of natural water flow with the geological effects of controlled water flow. In addition, natural versus controlled flooding can also be tested.

   (Recommended 20 pts.)

3. Conduct your research, recording your notes in the space provided below.

   Notes should include descriptions of a variety of the effects that dams have on rivers, including effects on the riverbanks and the ecosystem as a whole, as well as effects on the flow of sediment. Additionally, notes could include references to the effects of the Glen Canyon Dam on the Colorado River or other information from the CD-ROM articles.

   (Recommended 20 pts.)

Answer Key for Exploration 8, continued

4. Use the table below to record the effects of both types of water flow at varying flow rates.

Type of flow	Low	Medium	High
Natural flow	The stream's banks widen slightly, and small amounts of sediment are deposited on the inner curves of the stream.	The stream widens, and the banks become slightly more curved. A large amount of sediment is apparent.	The stream widens, and its banks curve much more sharply. Levels of sediment deposition continue to increase.
Regulated flow	The stream's banks widen slightly, just as they did during natural flow, but little or no sediment is deposited along the banks.	The stream widens, and the banks become slightly more curved, just as they did during natural flow, but little or no sediment is apparent.	The stream widens much more broadly during regulated flow than during natural flow. The curves are less snakelike because there is little or no sediment deposition.

(Recommended 30 pts.)

5. Compare the results of your stream-table observations with the information you discovered in your research. Which environmental and geological effects of dams are not reproducible by a stream table?

   Several effects of the dam are not represented by the stream table. Examples include the long-term effects of the loss of sediment that the river naturally carries downstream, the temperature difference in the water caused by controlled flooding, and the severe effects on the ecosystem as a whole.

   (Recommended 15 pts.)

Record your conclusions in the fax to Ms. Banks.

EXPLORATION 8 • ANSWER KEY

Answer Key for Exploration 8, continued

Exploration 8
Fax Form

FAX

To: Ms. Sandy Banks (FAX 520-555-7239)

From:

Date:

Subject: Possible effects of a dam on the natural river environment

What effect does a dam have on the natural river environment?

Because a dam blocks the natural flow of a river, it limits the erosion and deposition of sediment by the river. So instead of collecting and carrying nutrient-rich sediment downstream, the river drops the sediment at the bottom of the reservoir. As a result, the ecosystems downstream are deprived of the sediment that is often the basis of their food chains. Dams also prevent flood waters from periodically scouring the river bottom. As a result, debris settles on the river's bottom and creates barriers in front of inlets where fish would normally spawn. Where flood waters once carved out the river's banks and created new sandbanks farther downstream, vegetation takes root. These changes in the landscape threaten the organisms that live in and along the river system. *(Recommended 25 pts.)*

Answer Key for Exploration 8, continued

Which volume of water flow has the greatest impact on the formation of geologic features?

| Low | Medium | **High** |

Is it possible to maintain a healthy river environment downstream from a dam?

| YES | **NO** |

(Recommended scoring for the above two questions: 50 points for two correct answers; 30 points for 1 correct answer; 0 points for no correct answers.)

Why or why not?

Dams are very disruptive to the natural river environment. Even if water could flow through the dam at the same rate it would normally flow, much of the river's sediment would remain on the bottom of the reservoir. Controlled flooding could help keep the river clear of debris and could help distribute some of the sediment below the dam, but sediment above the dam would still need to be moved somehow from behind the dam to areas downstream. Also, some method would need to be devised to increase the temperature of the water before it affected the ecosystems downstream.

(Recommended 25 pts.)